To Carole –
for her patience in the
preparation of this book

CAMBRIDGE AEROSPACE SERIES

Aircraft noise

MICHAEL J. T. SMITH

Chief of Powerplant Technology (Civil)
Rolls-Royce plc

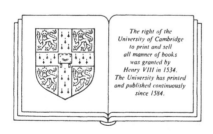

The right of the
University of Cambridge
to print and sell
all manner of books
was granted by
Henry VIII in 1534.
The University has printed
and published continuously
since 1584.

CAMBRIDGE UNIVERSITY PRESS

Cambridge
New York   Port Chester
Melbourne   Sydney

PUBLISHED BY THE PRESS SYNDICATE OF THE UNIVERSITY OF CAMBRIDGE
The Pitt Building, Trumpington Street, Cambridge, United Kingdom

CAMBRIDGE UNIVERSITY PRESS
The Edinburgh Building, Cambridge CB2 2RU, UK
40 West 20th Street, New York NY 10011–4211, USA
477 Williamstown Road, Port Melbourne, VIC 3207, Australia
Ruiz de Alarcón 13, 28014 Madrid, Spain
Dock House, The Waterfront, Cape Town 8001, South Africa

http://www.cambridge.org

© Cambridge University Press 1989

First published 1989
First paperback edition 2004

A catalogue record for this book is available from the British Library

Library of Congress Cataloguing-in-Publication Data
Smith, Michael J. T.
Aircraft noise.
   p.   cm. – (Cambridge aerospace series)
Bibliography: p.
Includes index.
ISBN 0 521 33186 2 hardback
1. Airplanes – Noise.   I. Title.   II. Series.
TL671.65.S64 1989
629.132′3 – dc 19   89-30052

ISBN 0 521 33186 2 hardback
ISBN 0 521 61699 9 paperback

Contents

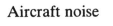
Aircraft noise

Preface

Most technological advances are accompanied by some degree of apprehension about their potentially catastrophic effect on safety and the environment. History provides countless examples. Notable in the field of transportation are the applications of the steam engine to the locomotive, the internal combustion engine to the horseless carriage, and the whole array of engines used in the flying machine. Immediate public safety is usually the primary concern, but pollution and noise are also cited as possible sources of long-term harm to the human species. Although the passage of time has seen some of these concerns justified, humanity has learned to live with them, whilst exploiting to the full the benefits of a wide variety of transportation modes. Indeed, in developed nations, the quality of life would be downgraded considerably without the benefit of fast, reliable and diverse modes of transport. As a result, the cost of environmental protection is considered a necessary burden.

Noise, as an environmental issue, has had a major impact on both the conceptual and detailed design of only one source of power in the field of transportation, arguably the one that has had great impact on international relationships, the modern aircraft engine. It was in the early 1960s, following the successful introduction of the jet engine into commercial airline service, that aircraft noise became an issue of substance. Hitherto, it had been regarded merely as an aggravation, and grievances were settled either privately, by the parties concerned, or in the civil courts. Interestingly, in the United Kingdom this latter facility was regarded as unnecessary, and Parliament virtually prevented U.K. citizens from having recourse to it as far back as 1920. As the minister responsible for aviation policy at the time, the late Sir Winston Churchill decided that this move was appropriate as it would help to promote and expand air transportation.

After World War II, propeller-driven aircraft were the only mode of commercial air transport until the de Havilland Comet appeared in the spring of 1952. After a shaky start, the jet fleet finally "took off" with the emergence of the four-engined Boeing 707 and Douglas DC8 at the end of that decade. Their immediate success on the longer-distance routes led developers to use the jet in aircraft designed to operate over shorter routes. These emerged as two- and three-engined aircraft such as the European Sud Aviation Caravelle, de Havilland Trident, British Aircraft Corporation 1-11 and the much more successful Douglas DC9, Boeing 727 and 737 models from America. Well over 2000 jet-powered commercial aircraft were plying their trade by the late 1960s and, in the process, they outstripped the fleet of large propeller aircraft in total numbers, number of passengers carried and distances flown. At the same time, the jet became attractive to the business sector, with the result that the number of "executive" jets has generally matched those in the commercial fleet. Today the commercial jet fleet in the developed countries as a whole consists of over 8000 aircraft, which perform at least a million flights each month. Add to this the almost equal number of executive jets, albeit operating less frequently, and all the propeller-powered aircraft, and the frequency of operation at major airports becomes impressive. Sometimes, it is of the order of one every half minute.

It is easy to see why noise became an issue. The early jets were noisy – extremely so. Apart from heavy construction plant, unsuppressed motor-cycles and emergency service sirens, they were the loudest source of noise in the community. Moreover, the noise of the jet drew attention to the rapidly increasing numbers of aircraft. The noise and frequency of operation around major airports combined to catalyse public reaction in these areas. As a result, by the early 1960s, airport owners were forced to establish local noise limits, which aircraft were not permitted to exceed. Noise-monitoring systems were installed around the airport boundaries and in the communities to police the statutory limits. In some cases, noisy long-range aircraft were forced to increase their climb performance by departing "lightweight", with only sufficient fuel to take them to a less sensitive airport, where they could then take on the fuel needed to complete their mission.

Local airport action kept the situation under control for a period of perhaps four or five years, during which the spread of jet operations to the shorter routes demanded more widespread action. By the mid-1960s, governments in those countries most affected by airport noise started a dialogue that eventually led them to introduce the concept of noise certification. This occurred around 1970, just when the much quieter high-bypass turbofan appeared on the scene and made the task of certification compliance less of a burden than the industry was expecting it to be.

Noise certification refers to the process by which the aircraft manufacturer has to demonstrate that his product meets basic noise standards, in the same way that it has to meet safety standards, before it can enter commercial service. The main impact of noise certification was not felt until the late 1970s, when the requirements were made much more stringent and were more widely applied. Today all new civil transport aircraft, irrespective of type, shape, size, weight or design of power-plant, have to satisfy noise requirements in all the manufacturing states as well as in most of the nations that are signatories to the International Civil Aviation Organisation (ICAO) 1949 Chicago Convention on Civil Aviation. These noise certification requirements (for the control of source noise) are supplemented by a series of "operational" strictures, whereby individual nations prohibit the purchase and/or operation of older, noisier, types. Such strictures have already caused the premature demise of many models of aircraft like the DC8, B707, deH Trident and older Sud Caravelles, and restrictions now being proposed for the 1990s are likely to extend the inventory of early retirements. In addition, noise at the airport may be controlled through "local" action, such as night curfews, daytime scheduling, supplementary landing fees for noisy types, land-use control and monitored noise limits. In all, it has been estimated that about US$1 billion is added to the air transport industry's annual costs in pursuit of a quieter environment, through fundamental modifications to aircraft and their power-plant designs, the additional fuel needed to offset performance losses, maintenance and policy and administrative activities.

This book looks at the fundamental aircraft noise problem, examines special cases and provides the "student" of aeronautics (in the widest sense) with an insight into the technical issues surrounding this practical field of the application of acoustics. It concentrates on subsonic, fixed-wing commercial aircraft issues, the area of the business that caused the aircraft noise problem to develop so rapidly a quarter of a century ago. However, the book also touches upon other types of aircraft, and the References (cited by superscript indices in the text) allow the practising acoustician to obtain further details on most aspects of aircraft noise control.

Because this book is intended for those who have a general as well as a professional interest in aircraft noise, terminology that is in common use has been employed throughout and mathematical formulae have been kept to a minimum. Some of the idiosyncracies surrounding the abbreviations, acronyms and strict terminology are covered in the appendices.

Acknowledgments

I wish to thank the directors of Rolls-Royce plc for encouraging me to write this book and for allowing me to use information and artwork proprietary to the company, in respect of which they retain full rights-of-use.

Thanks are due also to the scientists and engineers at Rolls-Royce and the Royal Aircraft Establishment (RAE), England, whose work over the years has contributed in no small measure to the advanced state of the art in aircraft noise control reflected herein. The RAE, other authors and organisations, including the International Civil Aviation Organisation, Butterworths and Her Majesty's Stationery Office, are all thanked for granting permission to reproduce elements of their work.

This book is very much a view from one small corner of Europe and reflects my particular interest in the "jet" engine. Apologies are made in advance to those who may feel neglected as a result. Needless to say, the views expressed are my own and should not be taken to reflect the position of either Rolls-Royce or the authors referenced.

1

Human reaction to aircraft noise

Noise, litter, housing and road development, local lighting, crime and a miscellany of other factors all produce mixed emotional reactions in Homo sapiens. Noise has often been cited as the most undesirable feature of life in the urban community. Aircraft noise is second only to traffic noise in the city in its unsociable levels, frequency and time of occurrence, and is often at the top of the list in rural areas. The rapid spread of civil aviation, from the early "jet set" élite of the 1960s to the tourists of today, has only served to intensify the problem, in that the increased number of flights in and out of airports has given rise to much greater intrusion on community life and, hence, to "noise exposure".

The growth of the jet-powered fleet is outlined in Figure 1.1. Following the initial surge of purchases in the late 1950s and throughout the 1960s, fleet expansion steadied at about 5% per annum in the 1970s. Because aircraft have grown considerably in carrying capacity, the expansion rate since 1980 has declined to about 2–3%, but this rate is expected to be maintained to the end of this century.[30,46] Although, in general, aircraft have become progressively quieter since the introduction of the bypass and turbofan engines in the 1960s and 1970s, the reduction in the noise level has not sufficiently offset the increase in operations or their psychological impact to bring an end to the problem.

Although levels of noise "exposure" can be quantified in terms of physical variables, each person's reaction to noise depends on his or her tolerance level. This may not simply be the product of physical factors – such as absolute noise level, time and frequency at which the noises occur, how long they last, the "quality" or spectral composition of the noise – but, in the case in question, it is often related to a basic

Figure 1.1. History of the growth of the free-world commercial jet transport fleet.

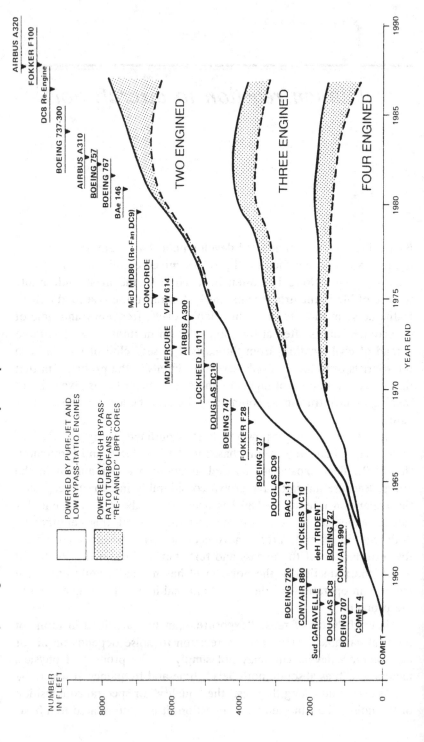

fear of aircraft. Nevertheless, for administrative and control purposes, all the real and imaginary effects that people perceive have to be rolled into a simple indicator; otherwise it would be far too complicated to quantify and judge the impact of changes to the general pattern of aviation, and the results might be misleading.

For this reason, rating aircraft noise has become something of an art.[111-21] To use a culinary metaphor, certain ingredients have been added to the basic stock, and the quantities varied to taste. Recently, however, authorities throughout the world have come to recognise the need for a standard recipe, or method of evaluation. Hence, it is important to stand back from all the specific issues to examine, firstly, the units and rating and exposure scales that have now been adopted universally.

1.1 Sound intensity, frequency and the decibel

Let us begin with some basic facts. "Noise" is merely common parlance for undesirable sound. Sound is a physical phenomenon, detected by the human being through one of five basic senses. Although sound-waves consist of only minute fluctuations in atmospheric pressure, their amplitude covers a wide range. Human beings have adapted to this phenomenon in such a way that they are capable of tolerating well over a millionfold change in the level of pressure fluctuation between the lower limit of audibility and the upper limit, which is marked by the onset of pain and may be followed by hearing damage. The unit of measurement used to compare the intensity of noises is the deciBel, which is more commonly written as decibel or dB. The decibel concept addresses the wide range of sound intensities accommodated by the human species in a similar manner to nature, by using a logarithmic ratio of the actual sound pressure level (SPL) to a nominal value, the "threshold" of hearing, set at 20 μPa. One Bel represents a tenfold increase in intensity, and the decibel a 1.26 change, that is,

$$[\text{antilog}_{10}(1/10)].$$

It follows that a doubling of sound intensity or noise level is reflected by a change of only 3 dB; the change in intensity (I) from I_1 to I_2 being expressed as

$$10\log_{10}(I_2/I_1) = 10\log_{10}(2/1) = 3.01 \text{ dB}.$$

This small (3-dB) numerical change for a doubling of sound energy is not generally appreciated; even less understood are the niceties of "perceived" noisiness. Where else can 50 added to 50 equal 53, not 100, and yet twice 50 be 60? The answers will emerge as we proceed.

The human hearing system not only tolerates a tremendous range of fluctuating sound pressure levels, but it also recognises an extremely wide range of frequencies. Frequencies as low as 20 Hertz (Hz), or cycles per second (cps), are sensed as sound. In young people, the audible range goes up to around 20 000 Hertz (20 kHz). Hearing deteriorates with age, and by midlife the upper frequency limit can be anywhere from 10 to 15 kHz. Even in the young, however, sounds above 10 kHz may have little loudness or annoyance effect because high frequencies are efficiently absorbed by the atmosphere and are of low intensity when received. Thus, for purposes of international standardisation, the upper limit of effective "noise" has been set at just over 11 kHz.

With such a wide range of frequencies to consider, it becomes desirable to subdivide them into more manageable ranges.[100-2] An octave band, which represents one-eighth of the defined audible range, encompasses a 2:1 spread of frequencies (e.g., 250–500 Hz). Because this is still a very wide frequency range to deal with analytically, the octave band is usually subdivided further into a minimum of $\frac{1}{3}$-octave bandwidths. The recommended or "preferred" octave and $\frac{1}{3}$-octave band frequencies established by the International Standards Organisation (ISO) are shown in Table 1.1.[102] The relationship between these, those bands used in defining aircraft noise and the more readily recognised piano keyboard is shown in Figure 1.2.

Because of the mixture of sound levels and frequencies involved, it is difficult to provide a standard definition of "noise level". To complicate matters even further, human beings do not enjoy a uniform response to sounds of the same intensity generated at different frequencies. The response of the human hearing system is such that it is most receptive to and, in the case of unwanted sound, most annoyed by that in the 2–4 kHz range (see Fig. 1.3). For this reason, any unit used to express human response to either loudness or annoyance has to include a weighting element that varies both with intensity and frequency.[103]

Over the years, many methods[104-10] of rating noise have been proposed, with the accepted scales being developed on the basis of human reaction to "loudness".[123-6] The widely used A-weighted (dBA) scale is a prime example, and is conveniently available for direct read-out on commercial sound-level meters via an electronic weighting network. Unfortunately, such meters are not normally useful in judging aircraft noise, which is unique in that it has a wide-ranging and variable spectral character and a transient, or rising then falling, intensity–time relationship. For this reason, special assessment scales[127–33] have been developed, which are based upon annoyance rather than loudness and which contain factors that account for special spectral characteristics and the persistence of the sound. Most other units, like the dBA, have been developed to assess the more general impact of industrial or other continuous everyday sounds, and are associated with loudness. The perceived noise[132] and effective perceived noise[133] scales (PNdB and EPNdB) are uniquely related to

Table 1.1. *Table of ISO preferred frequencies (Hz)*

Lower limiting frequency		Preferred centre frequency		Upper limiting frequency	
*	22.4		25		28.2
	28.2	*	31.5		35.5
	35.5		40	*	44.7
*	44.7		50		56.2
	56.2	*	63		70.8
	70.8		80	*	89.1
*	89.1		100		112.2
	112.2	*	125		141.3
	141.3		160	*	177.8
*	177.8		200		223.9
	223.9	*	250		281.8
	281.8		315	*	354.8
*	354.8		400		446.7
	446.7	*	500		562.3
	562.3		630	*	707.9
*	707.9		800		891.3
	891.3	*	1 000		1 122
	1 122		1 250	*	1 413
*	1 413		1 600		1 778
	1 778	*	2 000		2 239
	2 239		2 500	*	2 818
*	2 818		3 150		3 548
	3 548	*	4 000		4 467
	4 467		5 000	*	5 623
*	5 623		6 300		7 079
	7 079	*	8 000		8 913
	8 913		10 000	*	11 220
*	11 220		12 500		14 125
	14 125	*	16 000		17 783
	17 783		20 000	*	22 387

Note: Asterisks denote octave band values.

aircraft annoyance. However, the EPNL is a complex unit and, despite being based on loudness, the dBA is now often used to measure aircraft noise – for example, in the noise certification of propeller-driven light aircraft, in plotting noise contours and in setting local airport restrictions, where it is important to keep noise monitoring as simple as possible and provide a direct read-across to other sounds and

Figure 1.2. The audible frequency range used to describe aircraft noise.

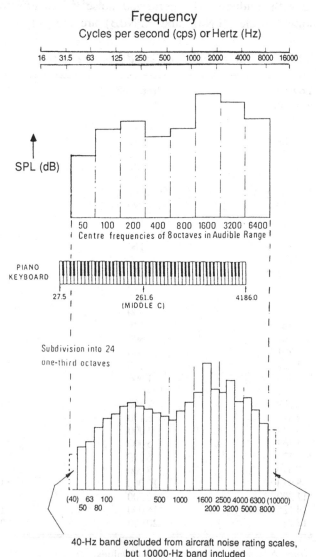

40-Hz band excluded from aircraft noise rating scales, but 10000-Hz band included

circumstances. Nevertheless, the PNdB and EPNdB are the corner-stones of technical and legal evaluations, as they are used in aircraft noise certification, and their derivation needs to be understood.

1.2   The perceived noise scale

In the 1950s, the "annoyance" response curve associated with the human hearing system, typified by Figure 1.3, was quantified by means of audiometric tests that covered a range of discrete (single-frequency) and broadband (multifrequency) spectral characteristics, including real aircraft sounds, to provide the basis for the first PNdB scale. Unfortunately, this first quantification was hastily constructed and proved short-sighted, in that it used only young college students as the test subjects. Their acute hearing and better-than-average high-frequency response led to overemphasis in the rating scale. Upon subsequent reevaluation,[130] using a more representative cross-section of the population, the high-frequency weighting was reduced to the level that now appears in the international PNdB scale. Figure 1.4 shows how the average observed response of the test audience was expressed in terms of both level and frequency, and identified numerically (in terms of a unit somewhat humorously designated the "Noy"), the reference being sound of random nature* in the 1-kHz $\frac{1}{3}$-octave band. A fairly simple mathematical transposition was then developed that integrates the Noy values over an analysed spectrum (either on an octave or a $\frac{1}{3}$-octave basis) to give a single-number annoyance value in perceived noise decibels (PNdB). Appendix 4 describes the procedure for calculating PNdB, whilst Figure 1.5 gives

Figure 1.3. Human response to sounds of constant intensity across the audible range (1 kHz reference).

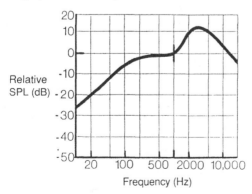

* A sound with a random variation of sound pressure with time.

an indication of how aircraft noise relates to some other everyday sounds, on the PNdB scale.

The PNdB unit and its associated level scale (PNL) make it possible to classify steady-state sound in terms of annoyance. Because the human aural system is sensitive to mid to high frequencies, the same

Figure 1.4. Contours of equal noisiness (Noy values).

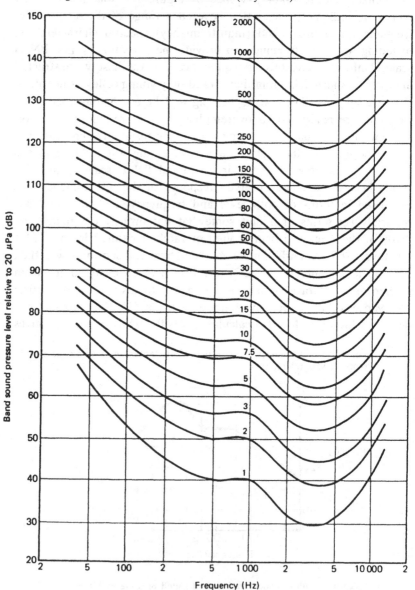

physical overall sound pressure level (OASPL) can result in a variation of up to 20 PNdB as the frequency distribution is altered. Since the PNL scale indicates that a change of 10 PNdB reflects a halving (or doubling) of annoyance, a 20-PNdB change reflects a fourfold variation in annoyance response. This is illustrated in Figure 1.6, where a

Figure 1.5. Typical everyday noise levels on the PNdB scale.

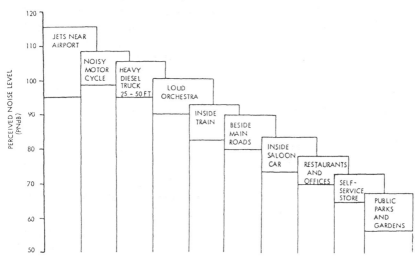

Figure 1.6. Computed perceived noise level of sounds of equal intensity and similar spectrum shape, but peaking at different frequencies in the audible range.

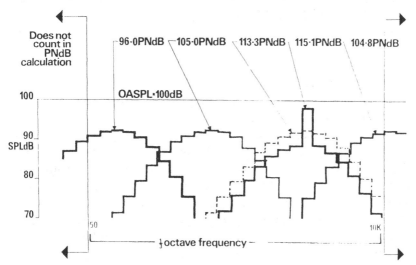

constant $\frac{1}{3}$-octave spectrum shape (with the same OASPL) has been moved through the audible frequency range. The resulting range of PNdB values is 17.3 dB. Furthermore, in one special case the PNL value is even higher, even though a slightly lower general spectrum level has been plotted. This is because one of the $\frac{1}{3}$-octave bands has been set at an augmented value, to give the same OASPL as the other spectra. This allows us to simulate the effect of a prominent discrete tone falling within the bandwidth of a single $\frac{1}{3}$-octave, the reason being that the human hearing system is able to detect and preferentially "lock on" to a discrete tone and, in the process, will tend to disregard sound of a more broadband nature in the adjacent frequency range. As a consequence, the discrete tone is often perceived as being more annoying than the plain PNL scale would indicate. Consider how effective the modern emergency service sirens have become!

1.3   The tone correction

The tone-sensing characteristic of our hearing system is yet another factor that has to be considered in judging the annoyance of an aircraft.[135-8] The scale used to describe this phenomenon is the tone-corrected perceived noise level (PNLT). In this scale, a special "penalty" is added to the perceived noise level, which varies with both the degree of intrusion of the tone (above the local ambient level) and its frequency. This procedure is said to take full account of the effects of protrusive discrete tones within the matrix of sound levels and frequencies detected by the human being, although no new work has been done on this subject since the original analyses were performed in the 1950s and 1960s. The currently accepted exchange rate of annoyance with frequency and level is illustrated in Figure 1.7 (the calculation procedure for PNLT is described in Appendix 4). The fact that only two correction curves are provided reflects the gross nature of the assumptions made about the influence of tones, whilst the fact that the curves change slope near the origin illustrates the influence of "technopoliticking" (see Chapter 6).

Originally, the recommended correction[133] recognised that an attempt was being made to allow only for "protrusive" tones and that the experimental evidence was thin. Accordingly, no attempt was made to correct for tones that did not protrude at least 3 dB above the local $\frac{1}{3}$-octave background level. This produced a cut-off, or "fence", at the 3-dB protrusion, which stood the test of time and was an integral feature of the noise certification process until the mid-1970s. Then, with the best of intentions, one particularly tidy-minded delegation to the International Civil Aviation Organisation persuaded members to "clean up" the bottom end of the curves by extending them all the way down to the origin.

Fortunately, this disastrous and quite arbitrary modification was comparatively short-lived. Even so, it was three years before all concerned were convinced that the new correction deemed even a broadband sound source to be riddled with "protrusive" tones, after it had

Figure 1.7. The tone correction factor of ICAO Annex 16.

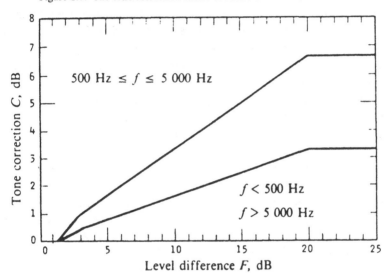

Frequency f, Hz	Level difference F, dB	Tone correction C, dB
$50 \leq f < 500$	$1\frac{1}{2}* \leq F < 3$	$F/3 - \frac{1}{2}$
	$3 \leq F < 20$	$F/6$
	$20 \leq F$	$3\frac{1}{3}$
$500 \leq f \leq 5\ 000$	$1\frac{1}{2}* \leq F < 3$	$2F/3 - 1$
	$3 \leq F < 20$	$F/3$
	$20 \leq F$	$6\frac{2}{3}$
$5\ 000 < f \leq 10\ 000$	$1\frac{1}{2}* \leq F < 3$	$F/3 - \frac{1}{2}$
	$3 \leq F < 20$	$F/6$
	$20 \leq F$	$3\frac{1}{3}$

been subjected to the normal processes of atmospheric propagation and ground reflection. Nevertheless, somewhat characteristically, the "technopolitical" mind could never accept that it had simply made a mistake and the compromise solution of Figure 1.7 was only grudgingly agreed upon as a face-saving expedient in the early 1980s.

As regards the differentiation between tones inside and outside the 500–5000-Hz band, it would be all too easy to overemphasise the impact of very high frequencies, as the original PNdB scale did. Also, high frequencies suffer large attenuation via natural atmospheric absorption. Hence, the lesser emphasis on tones above 5000 Hz is understandable and justified. However, tones below 500 Hz create a problem. The fact that they frequently protrude well above the general spectrum level is undisputed – propeller-powered aircraft and helicopters can easily have tone protrusions of 30 dB. But, should they be heavily penalised or does the basic PNdB scale adequately reflect their impact without requiring a further penalty? The debate continues to this day. It has been suggested that the $\frac{1}{3}$-octave subdivision is too broad at very low frequencies, in that it is incompatible with the way in which the human hearing system performs, and that the bottom nine bands should be grouped.[131] It has also been claimed that the tone penalty at frequencies below 500 Hz can simply be ignored.[166–9] No doubt, the debate will be given added impetus[170–1] if the new "open-rotor" (advanced propeller) power-plants currently being developed ever become a production reality.[169–71]

1.4   The effective perceived noise scale

The tone-corrected unit is an important ingredient in the noise descriptor now accepted internationally in the formal aircraft noise certification process, the effective perceived noise level,[133] or EPNL (in units of EPNdB). This scale is further complicated in that it contains yet another ingredient specific to aircraft noise, a "duration" correction, which recognises the rising and falling nature of aircraft noise, and the fact that duration varies with both the type of aircraft and the mode of operation.[139] As illustrated in Figure 1.8, the nominal duration of an aircraft noise–time history is defined on the basis of the time during which the sound is at a level above a threshold 10 dB below the peak (tone-corrected) value. In the standard method of computation, the energy below the PNLT curve and above the 10-dB-down "threshold" is fully integrated and normalised with respect to a time constant of 10 sec. In practice, a typical take-off noise signature at the certification reference point, 6.5 km from brake release on the runway, will have a duration of some 10–20 sec, leading to a positive duration correction; whereas on approach to landing the correction is always negative, since the duration is about half of the standard 10 sec.

With this final correction, we arrive at the effective perceived noise level, a complexly derived single-number expression of the annoyance caused by the noise from a single aircraft operation, as experienced at one position on the ground.

The derivation of an EPNL value from flight testing is closely controlled by international standardisation through the noise certification specifications of ICAO Annex 16. Approved values for all major aircraft are published in flight manuals by the relevant airworthiness (certification) authority, and a listing is periodically compiled by ICAO. Nevertheless, not all nations use the information in the same way. Noise exposure rating is not quite as straightforward as determining the single-number EPNL annoyance value for one aircraft. This point needs further discussion.

1.5 Noise exposure modelling

Noise "exposure" is a euphemism for the calibration of the reaction of airport neighbours[141–3] to the operation of many aircraft in a repeatable time period, usually 12 or 24 hr. It expresses the effects of noise level, spectral character, duration of intrusion and less tangible environmental factors such as the local "ambient" background level,[139–40] which can vary over any given period of time. Over the years, most nations with an airport noise problem have developed their

Figure 1.8. The duration of an aircraft noise–time history.

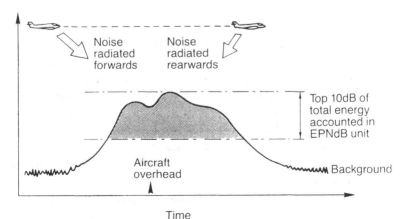

own annoyance descriptors and, as a result, noise exposure modelling has become a "fashionable" art, which in many cases is taken to extremes. The total operational pattern at a major airport may be modelled in the minutest detail and the noise pattern from all the individual aircraft operations over several hours integrated to build up an overall representation of the situation. The whole process can then be repeated to paint a picture of seasonal changes or trends over periods of months or years.

The first descriptor to be used on a national basis was the noise and number index (NNI),[144] which was developed in the United Kingdom during a major study into the effects of everyday noise on the population at large.[145] On the basis of a social survey of people living in the vicinity of London–Heathrow Airport[146] and a subjective-assessment experiment at the time of the 1961 Farnborough Air Show, the NNI attempts to relate annoyance to both the individual aircraft (peak) noise levels and the frequency of operations (noise intrusion), so as to produce a single-number annoyance or exposure indicator.

The NNI uses the PNdB as its basic noise unit, but adjusts the average of the peak PNdB levels that occur over a given time period by adding a correction factor equal to $15 \log_{10} N$, where N is the number of operations, to account for repeated intrusion. Other exposure indices, developed subsequently, use a lesser "mark-up" for the number of operations,[147–50] usually $10 \log_{10} N$. The reasons for the discrepancy are twofold. Firstly, the NNI was developed at a time when jet aircraft were a comparatively novel feature of everyday life in the United Kingdom, and public reaction reflected not only widespread dislike of the high levels of noise that these new machines produced, but also a fundamental objection to their appearance in the skies around London. As with all novelties, familiarity tends to breed contempt or, in this case, acceptance, as reflected in subsequent social surveys, which have shown less reaction to aircraft in large numbers.

The second factor was that the survey data on noise, on the one hand, and frequency of operations, on the other, were intended to be uncorrelated, but they were not. The sampling technique relied heavily on comments from people living close to the end of the runways at Heathrow, where all operations were noticed because the noise levels were high. Had the survey taken place in areas well away from the airport, where absolute levels were lower, there might well have been less reaction to frequent operations, as in the later surveys.[151]

Recent studies in the United Kingdom[152] have exposed the short-comings of the original work, and it is now recognised that the number of operations is less significant than inferred in the NNI studies. To its credit, however, the NNI has stood the test of time. It has been used consistently for a quarter-century to describe the annual noise trends around major U.K. airports. It has also been used to formulate legal guidelines, has set the criterion against which householders are able to claim financial support for sound-insulation work on their houses[16] and has established other demarcation boundaries between the acceptable and the unacceptable. On the debit side, however, the decision to apply a lower-level cut-off at 80 PNdB, on the basis that this value represents the threshold of annoyance, has possibly led to an understatement of the noise problem at small airports, where low operational noise levels are offset by a low local ambient.

The equivalent exposure measure used in the United States for many years was the noise exposure forecast (NEF), which has now been replaced by the day–night sound level (LDN). The NEF used the EPNdB as its descriptor, but the LDN has taken into account the growing use of the simpler dBA unit for airport noise contour calculations and in many monitoring systems. Although the dBA is a loudness-based unit and produces numerical values some 10–15 dB below the PNdB scale, it does have a much smoother response against frequency and is therefore far less susceptible to large changes for small variations in the source spectrum. In Figure 1.9, the dBD weighting, along with the nearly identical PNdB "N" weighting[134] recommended for sound-level meters by the Society of Automotive Engineers (SAE), are compared with the dBA. As is clear, the dBA weighting does have a different slope, broadly oscillating between dBD and a 15 dB reduced level, with very different high-frequency characteristics. However, high frequencies are rapidly absorbed by the atmosphere and, at the distances involved in evaluating the impact of airport noise, the effect of a change from PNdB to dBA in terms of relative annoyance is probably of little significance.

The LDN does, however, fully recognise the importance of the duration element by using the integrated dBA of the sound exposure level (SEL) rather than the basic instantaneous, or peak, dBA level (as might be measured with a simple sound-level meter). It also tries to adjust the SEL for time of day by adding a fairly arbitrary 10 dBA to reflect the increased annoyance when urban background noise levels are low, particularly during the late evening and overnight.[153–8] Although the amount of this allowance is based on shaky subjective support, it is simple and makes the LDN easy to compute, for full spectral knowledge of each aircraft movement is not required, as the time integration is based on the time-above-a-threshold level rather than full energy integration over the whole period of interest.

The actual timing, and the degree to which an adjustment should be made for lower background levels and lower noise intrusion at night has been a subject of considerable political and technical debate over the years. The LDN, in common with some other exposure indices, assumes a doubling of annoyance at night by the simple addition of a constant 10 dBA to all noises created in the night-time period. A few exposure indices make a graduated allowance as the evening progresses, although no community is going to respond in precisely the same way to aircraft noise at night – it depends on the nature of the community in question. In some city centres there is often more traffic noise than aircraft noise through to the small hours of the morning, whereas in rural communities the population may be safely tucked up

Figure 1.9. Weighting curves for sound-level meters.

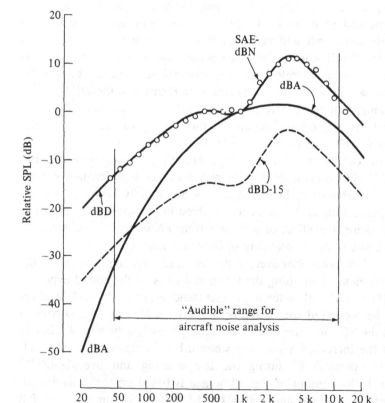

in bed at an early hour. Equally, the "siesta" period might be a critical time in some hot countries. All in all, however, there is undoubtedly a correlation between disturbance and ambient level/time of day. Perhaps the most pointed survey ever conducted (and there have been many over the years) was one that posed the following simple question to the communities around a major U.S. airport: "If you could stop the aircraft flying for one hour a day, which hour would that be?" The answers almost unanimously cited a late evening hour when people are trying to get to sleep. Once people are asleep, it takes a higher individual noise level from each aircraft movement to actually wake them up than it takes to stop them from going to sleep.

A major study on the effects of aircraft noise on sleep disturbance was undertaken in the United Kingdom[156] in the late 1970s. The objective of the programme was to establish the nature and extent of sleep disturbance close to major U.K. airports, from all causes, and to assess the relationship between aircraft noise and the disturbance. The exercise was conducted using social surveys in different areas that experienced a wide range of aircraft noise exposure, from one site at which no aircraft noise events were recorded to sites with levels as high as 106 dBA and, on average, forty aircraft movements per night. Night was defined as the period 2300–0700.

Aside from the main conclusion on sleep disturbance, the study found that another measure, the equivalent continuous sound level (L_{eq}),[159] gave a quite "satisfactory" measure of aircraft noise exposure, in that it correlated well with sleep disturbance. L_{eq} is an energy-averaging index, derived by summing the noise pattern over a given time and then normalising it to unit time. This can be achieved either by continuous energy integration or, where the nature of the noise pattern is intermittent, the period of recording may be compressed by applying a threshold-level "trigger" to the measurement system and then adjusting the average single-event sound level for the number of events, in the same manner as with other indices (see Fig. 1.10).

Although the U.K. study concluded that there would be technical advantages, L_{eq} has not yet been widely applied in the assessment of aircraft noise. It may or may not be just another noise index to add to the considerable list that has accumulated over many years, and it may be some time before a single index is used worldwide. What the future holds with regard to international cooperation is, of course, unknown. Progress to date has been slow, but there is now at least a measure of harmony in the methodology used to predict noise contours in that ICAO states are trying to bring together the common major elements of the several similar methods that exist.[990] To get all the nations concerned to agree on these elements will take some time, and it is to be hoped that things will not take so long that any agreed methodologies are out of date before they are widely used! It is important that the leading aviation nations have a unified approach to the use of exposure indices and that the one they finally select is trustworthy for a considerable period into the future.

The importance of credibility can be seen in the problem of human

Figure 1.10. Composition of the equivalent continuous sound level (L_{eq}).

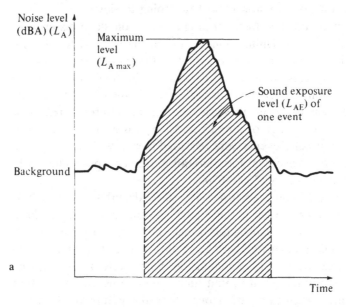

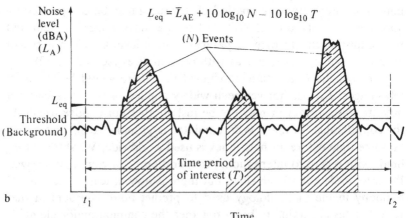

$$L_{eq} = \bar{L}_{AE} + 10 \log_{10} N - 10 \log_{10} T$$

reaction to the density of air traffic, and how reaction varies with absolute noise level. The emphasis placed on the "intrusion" factor – that is, on the effect of frequency of operation in and out of airports – is vital in determining whether modern aircraft are environmentally "acceptable" or whether there is going to be a continuing airport noise problem. Aircraft are getting quieter but, at any one time, there are many older, noisier types still in service which markedly colour the overall picture. When the world fleet is composed overwhelmingly of quieter aircraft, the noise intrusion on everyday life from each operation will be less, and the relationship between annoyance and the number of operations may also become far less significant in the overall noise exposure equation.

Perhaps we have already reached the point where the number-related correction factor used in most indices should be deemphasised further but, unfortunately, it often takes far too long for an established scale to be modified to reflect a changing situation. Consequently, many forecasts are out of date and misleading almost as soon as they are published. "Administration", that is, national or international bureaucracy, has a habit of dragging its feet and, in the process, important decisions are taken belatedly. The importance of individual elements in any scale is discussed in Chapter 8; it suffices here to say that there is no fully reliable scale in common usage today and that there is still a great need for further research into human reaction to aircraft noise. Many of the "established" elements require closer examination in the light of a quarter-century of experience with the airport noise problem.

2

Action against aircraft noise

The noise from the early pure jet and low-bypass turbofans was, in hindsight, wholly unreasonable. The public should never again be permitted to suffer such intrusion on sleep, conversation and relaxation without adequate redress via compensation from the aircraft suppliers and operators, or the customers who benefit from the service provided. Air travel has penetrated international frontiers, brought nations closer together and boosted world markets but, in the main, it *is* a luxury that pampers the business community and the tourist on a daily basis. Without air transportation, many people would not have the pleasure of visiting the numerous faraway places that are now readily accessible to them in a comparatively short travelling time. Similarly, on the commercial front, mail and freight services would be slower and the wide variety of perishable goods currently available would no longer be able to reach world markets. Business travel, which frequently represents the backbone of the revenue from passenger operations, would be reduced to the essential minimum, but the world would still manage, perhaps at a less frantic pace!

In order to enjoy the many benefits of air transportation, many people have had to suffer high levels of noise intrusion, in most cases, not of their own choice. Airports have expanded, many new ones have appeared, and "noise-impacted zones" have spread to embrace areas that have traditionally enjoyed an element of serenity. Although few people have seriously suggested that aircraft noise is detrimental to health,[164] the impact of noise in general,[105-10] and of aircraft noise in particular,[160-3] has been, and still is, the subject of wide-ranging debate. Opinions vary, but it cannot be disputed that governments must accept a good deal of the blame for the problem, for they saw it coming as they encouraged the development of the air transport industry through military and civil R&D funding, yet they repeatedly ignored the pleas of the innocent bystander. In some cases, instead of controlling the situation, they made it worse. For example, in 1920, the late Sir Winston Churchill legislated away the right of U.K. citizens to sue for the nuisance created by the noise of aircraft in flight so that the spread of aviation could be encouraged. In 1947 this was extended to cover the noise of aircraft on the ground, and in 1960 aircraft noise was specifically excluded from the Noise Abatement Act. More generally, all the early civil jets were the offspring of military R&D funding programmes, and the airlines readily seized the opportunity to expand markets by buying jets in large numbers. The manufacturers' initial attitude to the noise problem was that the power-plants were "legal" and that noise was an environmental, not a commercial, issue.

Belatedly, governments did act. In the late 1960s, after four years of prevarication, they introduced legal strictures on the manufacturing industry (not the airlines) that demanded minimum noise standards before an aircraft could enter service. The events leading up to this situation, and subsequent actions, are interesting in their own right.

2.1 History

The de Havilland Comet was the earliest commercial aircraft to be powered by a jet engine. The original prototype first flew forty years ago (i.e., in July 1949) and, following proving tests on a second prototype, the first production airframe entered service on 2 May 1952. Sadly, the Comet's potential was blighted by a series of high-altitude structural failures, which curtailed operations and led to the redesigned and longer-range Comet 4, which first flew in April 1958. Less than 130 Comet airframes were produced, including those later operated by the British military as maritime Nimrods, and the installed thrust was only some 10–20% of that required for a modern long-range aircraft.

The noise created by the early military jet engines forewarned the industry of a problem, and the reaction to the civil de Havilland Ghost and Rolls-Royce Avon engines in the Comet was sufficient to confirm a need for extensive work on the noise of the gas turbine aeroengine. This need was made even clearer by the public's reaction to the other major commercial jet-powered aircraft. The long-range Boeing 707, which was first delivered in December 1958, its derivatives and the Douglas DC8 of 1962 reached production figures of over 1500 before they were superseded in the early 1970s. Operation of the B707, in particular, was a major environmental and political issue during its early years of service. The noise created by the Pratt and Whitney JT3 and JT4 and the Rolls-Royce Conway engines was sufficient to force the airport authorities at London Heathrow and New York Idlewild (now Kennedy) airports to institute noise limits and install monitoring systems to police their observance. Long-range operations that failed to meet the limits were forced to initiate their service at lightweight (to climb away faster and get further from the community quickly) and to take on their full fuel requirements at intermediate airports, such as Shannon and Prestwick. Nevertheless, noise monitoring and policing of imposed limits failed to stem the problem, or to prevent the development of a whole series of aircraft that exploited the commercial opportunities offered by the speed and general passenger attraction of the jet.

The Sud Aviation Caravelle emerged for shorter routes after further development of the Avon engine, which had originally powered the Comet, and in the mid-1960s both Douglas and Boeing produced short- to medium-range aircraft with the DC9, B727 and 737, all of which used the most numerous commercial engine ever produced, the Pratt and Whitney JT8D. In Europe, de Havilland and Vickers/British Aircraft Corporation supplemented the fleet with the Trident, VC10 and 1-11. In fact, well over 2000 jet-powered commercial aircraft were in service by the late 1960s and, by the time positive government action was taken, the world jet fleet well outnumbered the established airline propeller-powered fleet.

It was in 1966, following a significant number of lawsuits in the United States and the public outcry in major European cities, that positive action was initiated. An international conference was called in London, which endorsed the concept of manufacturer control by "certification". In the United States, the associate administrator for development in the Federal Aviation Agency (FAA) wrote to selected U.S. manufacturers warning them of proposed federal action:[4]

> The growth and public use of aviation is most encouraging. However, it could well be that the most significant deterrent to continued growth is related to problems associated with aircraft noise. Noise has become a problem of national concern in our society. People are becoming increasingly disenchanted with illusory statements to the effect that increasing noise levels are synonymous with industrial progress. We know that you are aware of continuing complaints from the public and mounting congressional interest in solving the problem. Recognizing this, the President, in his message to Congress in March 1966, enjoined the Administrators of the Federal Avia-

tion Agency (FAA) and the National Aeronautics and Space Administration (NASA), together with the Secretaries of the Department of Commerce (DOC) and the Department of Housing and Urban Development (HUD), to work closely with the President's Science Advisor in the Office of Science and Technology (OST) to organize a broad national program to insure coordinated federal effort in seeking solutions to this most vexing problem. He further directed them to study the development of aircraft noise standards and recommend legislative or administrative actions needed to move ahead in the area of noise relief.

In consonance with this, the FAA has drafted legislation which has been introduced in both the Senate and the House of Representatives to amend the Federal Aviation Act of 1958 to provide specific authority to regulate in the area of aircraft noise. The language of the legislation means, assuming it is enacted by the Congress, that the Agency will require compliance with noise standards as well as compliance with safety standards as a condition to the issuance of future type certifications.

The letter then discussed the proposed framework for new legislation, and identified the selection of the special noise descriptor (EPNdB), closing,

In summary, we believe that aircraft noise levels have assumed an unprecedented unpopularity with large segments of the public. The accelerated growth of air transportation combined with the development of larger and more powerful aircraft, the enlargement of many airport facilities, and the engulfing of lands immediately surrounding our airports by urban communities all contribute to a growth of the noise problem on a national scale. We believe the Agency must act in a positive manner to check this growth if we are to see aviation continue to serve our Country's needs.

The FAA then moved on to consider practical structures for a noise certification scheme, which it formally announced to the public in 1969 via a Notice of Proposed Rule Making (NPRM 69-1). Meanwhile, the same FAA personnel who devised NPRM 69-1 had been representing the U.S. government in discussions with the United Kingdom and France concerning the possibility of an international scheme and, in 1969, a special international meeting was held under the auspices of

ICAO. This resulted in the creation of the ICAO Committee on Aircraft Noise (CAN), which was charged with taking positive and urgent action.

The publication of NPRM 69-1 in the United States, before ICAO had come to any agreed solution to the escalating airport noise problem worldwide, reflected U.S. impatience with international politicking. It also spurred ICAO into action earlier than might otherwise have happened.

By 1971, the voluminous public and industry responses to NPRM 69-1 had been thoroughly sifted by the FAA, and the first noise certification scheme emerged as U.S. domestic legislation in the form of FAR Part 36,[5] a new section in the Federal Aviation Regulations. FAR Part 36 became retroactive to the date of the original NPRM of January 1969, and aircraft that had been manufactured in the intervening period were given a breathing space in which to comply. Since the manufacturers of the L1011 and DC10 had been fully aware of these developments and had designed both aircraft to meet the proposed rules, the only aircraft caught retrospectively were the first 150 Boeing 747s. These required engineering modifications to the power-plant inlets, which added about 1% to the cost of each aircraft.

2.2   The noise certification scheme

The international body eventually caught up with the United States, and in 1971 the recommendations of CAN were published by ICAO[1] in the form of Annex 16, a formal addendum to the proceedings of the 1944 Chicago Convention on Civil Aviation. However, unlike the FAA's decisions (which have the power of U.S. law behind them), ICAO's decisions have to be accepted by each member state and written into their legal frameworks. It is only after this process has taken place that the local certificating authority[5,7] can implement them. It often takes over two years before there is widespread acceptance.

As to the details of the rules, although the frameworks of both the original FAR Part 36 and ICAO Annex 16 were similar and, in fact, identical in principle, they were sufficiently different in several critical details to make the U.S. regulation more stringent. Both were directed at imposing a ceiling on the noise permitted by commercial aircraft and did so by specifying limits at the same three critical operating conditions around the airport. As shown in Figure 2.1, the three conditions relate to noise under the take-off and approach flight paths and noise to the side of the runway (the only condition in which the engines are always operating at full power). The location of these three points was devised with one eye on population dispersion around airports, and another on establishing a single "never-to-exceed" number, which would serve to enclose the airport in a noise "box". The number agreed on was 108 EPNdB, the effective perceived noise scale having been created specifically for aircraft noise certification purposes (see Chapter 1). The value of 108 EPNdB was some 6–10 dB below the maximum noise generated by the existing long-range category B707, DC8 and VC10 aircraft and hence represented a 25–50% reduction in annoyance as perceived by the population.

Since the short- and medium-range sectors of the fleet were by no means silent, they also merited attention in controlling the total airport noise problem. This was achieved by establishing a sliding scale of noise versus take-off weight, up to the maximum 108 EPNdB. The actual limits imposed in both FAR Part 36 and Annex 16 are shown in Figure 2.2. The rationale for the slopes was purely empirical, and the lower-weight constant limits reflected noise levels compatible with small business and general aviation aircraft, in effect, those below which nobody registered particular concern at major airports. This is not to say that they could be described as "acceptable" levels.

The differences between the U.S. and ICAO requirements lay in slight variations in the distances to the three reference measurement points, and in the power settings and flight speeds permitted during compliance demonstration. These had the general effect of making FAR Part 36 more stringent than the ICAO regulations, by as much as

Figure 2.1. The three noise certification reference positions.

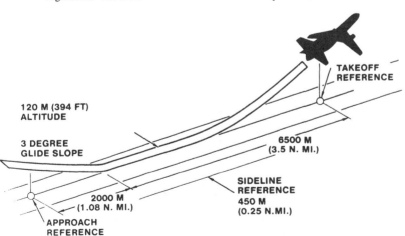

5 EPNdB, and prudent manufacturers attempted to meet the American rules to ensure worldwide acceptability. Fortunately, today the U.S. and ICAO rules are almost identical, and confusion is avoided.

As implied earlier, the only aircraft that were truly designed to meet these new regulations in the early years of their applicability were the high-bypass-powered "wide-bodies" that followed that notable first,

Figure 2.2. Relative severity of ICAO Annex 16 and FAR Part 36 original noise limits.

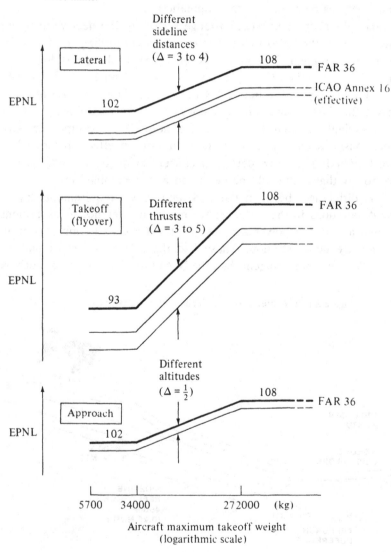

the Boeing 747 jumbo jet. This was the only aircraft to have difficulty in meeting FAR Part 36, partly because it was designed before the FAA became serious about noise certification, but also because it was well outside commercial aviation experience. It was twice the size of anything that had preceded it, had totally new power-plant designs, and suffered by being up against the ceiling at the top of the noise-versus-weight scale.

The other major airframe manufacturers (Lockheed, Douglas and Airbus Industrie) and their engine suppliers all had the benefit of advance warning of the new rules and invested heavily in noise research and development. As a result, the L1011, DC10 and A300 had power-plants with more built-in noise-control features than the early B747s, and all met the requirements handsomely. However, these successes convinced politicians that it would be possible to prescribe even lower limits for aircraft produced by modern technology and so demonstrate to the public the environmental "awareness" of government. Even so, nothing happened on this front for another six to seven years.

2.3   Certification develops

The next stage of legislation was, in fact, prompted by the realisation that FAR Part 36 and Annex 16 were having little overall impact on the noise problem at the airport. The world fleet was dominated by aircraft designed and manufactured in the 1960s, with engines some 20 dB noisier for the same thrust than those installed in the emerging fleet of wide-bodies. There were no high-bypass-ratio turbofan engines being produced to power the enormous fleet of aircraft that catered for airline seating requirements of less than 250. These requirements were still being filled by a range of aircraft powered by the Pratt and Whitney JT3D and 8D engines, and to a lesser extent by smaller European engines, such as the Rolls-Royce Spey. In particular, Boeing's 727 and 737 models were selling extremely well, as were the various versions of the Douglas DC9. Certification did not constrain the continued production of these aircraft, and since, in the words of the FAA, their JT8D engine *was* "the U.S. airport noise problem", the eyes of the administrators turned towards embracing this type of aircraft in the certification net. In the period 1974–7, firstly the United States and then ICAO extended the provisions of the existing certification requirements to cover every individual new airframe that came off the production line. This action led to an enormous flurry of activity in the industry, with "hushkits" being developed for every aircraft that could not comply with the certification requirements. For some, like the B707, this move spelled the end of the production line, for they could not be suppressed economically.

In simple numerical terms, those hushkits that reached the production stage reduced noise levels by as little as 1 EPNdB and at most by 3–4 EPNdB. But this was all that was required to make the relevant aircraft "legal" and, as a consequence, the whole exercise was largely cosmetic, despite being extremely expensive. A change of 3 EPNdB is virtually undetectable by the human ear, with 10 EPNdB being necessary to reflect either a halving or doubling of annoyance. However, this did not deter those who represented their image-conscious political bosses, even though a wide range of noisy aircraft became the proud owners of an environmental "seal of approval". On the credit side, however, the flurry of activity that the hushkit game created did make the industry realise that many governments around the world were happy to make the aircraft noise problem a political football.

This fact was driven home during the next stage of the legal process, for it did not take long for the "system" to realise that the application of new-type standards to a limited production line of very large aircraft, coupled with the imposition of demands for only marginal noise reductions in the rest of the fleet, was having very little impact on the overall pattern of noise exposure around airports. It was just possible that some operationally "saturated" airports were seeing a small benefit from the endeavours of governments worldwide, but those living near an expanding airport were in no position to thank their elected representatives. Accordingly, the technology that the industry had now demonstrated to be available was demanded in extended legislation. Those provisions of Annex 16 that were applicable to completely new designs of aircraft were made considerably more stringent in the 1975–8 timescale. This followed an extensive international debate on whether the certification scheme was sensibly framed and what level of increased stringency was appropriate. On the former issue, ICAO established working groups to look at the suitability of noise certification methodologies for different classes of aircraft (jet-powered, propeller-powered, supersonic transports, etc.) and on the latter issue it took stock of the state of noise technology throughout the industry. Unfortunately, ICAO was forced to make early changes to the legislation to prevent newer types of aircraft from "growing" up to the noise limits already established and made rather a mess of things in the process. At its fourth meeting in 1975, ICAO–CAN established tougher limits for the new types of jet-powered aircraft by the simple expedient of lowering the EPNL limits already established. This process was far from equable but, since no firm recommendations had come from its working groups on more appropriate methodologies, ICAO had to do something so that it would be seen to be reacting to the growing demands for action. Fortunately, although formally published and accepted by all member nations, these new regulations were never actually put into effect, since no new aircraft designs emerged between 1975 and 1977. By the fifth meeting of CAN in 1977, the working groups had studied several alternatives to the inflexible methodology pioneered by the United States in 1969. However, they were unable to make firm recommendations concerning any substantial changes owing to the intransigence of some members, notably the United States. There had been a groundswell of opinion,[27-8] supported by extensive French and U.K. parametric studies, in favour of changing the system to reflect airfield performance rather than aircraft weight. In this way, each aircraft could be judged according to its likely airfield usage, and hence could address the real airport situation, rather than the artificial certification "airfield". However, the necessary changes to the certification format were not accepted, as ICAO voted for a compromise between flexibility and rigidity. The compromise, originating in the United States, introduced the somewhat strange concept of setting different take-off noise requirements for aircraft with different numbers of engines. This philosophy, illustrated in Figure 2.3, reflects the fact that the airworthiness requirements demand that an aircraft is capable of taking off even if it suffers an engine failure during the most critical phase of take-off, that is, after there is insufficient runway left in which to brake to a standstill. This requirement means that a twin-engined aircraft has 100% more thrust than the minimum it requires under normal circumstances and, with both engines operating, it climbs extremely quickly. By comparison, a four-engined aircraft has only 33% excess thrust and with all engines operating climbs substantially less steeply. The trijet falls between the two, with 50% excess thrust.

Hence, the four-engined aeroplane is comparatively close to the community when it overflies the reference microphone at 6.5 km from brake release; the trijet is somewhat higher and the twin even higher still. The greater the distance between the community and the aeroplane, the quieter it appears and, therefore, the same engine installed in two-, three- and four-engined aircraft will give substantially different community levels. Consequently, the two-, three-, and four-engine rule was devised on the basis of "equal agony" in engine noise-control technology. The flaw lies in the practicalities of everyday operation. Why should a three-engined DC10 have to make less noise than a four-engined B747 that operates alongside it from the same airports and over the same routes? Or why should a BAe146 be permitted to make more noise than a B737–300 or Fokker 100?

Despite such anomalies, the latest requirements of Chapter 3 of the 1977–8 version of Annex 16 and the U.S. equivalent, FAR Part 36 Stage 3, have stood the test of time, and remain unaltered to this day (see Appendix 6). This is primarily because more than 50% of the world fleet is still composed of aircraft that were only required to meet the initial certification requirements of 1969, and because the industry has not demonstrated that it is capable of bettering the latest requirements by a margin big enough to justify administrative action. What has happened since 1977–8, is that certification requirements have been extended to cover just about every other class of aircraft that

Figure 2.3. Derivation of ICAO Annex 16, Chapter 3, take-off noise limits.

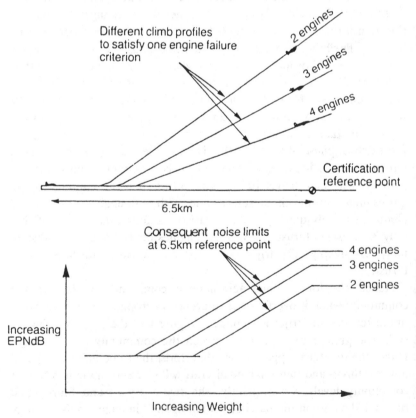

exists, and the various stages of jet aircraft legislation have been used as demarcation lines for discrimination in everyday operation. For example, at some airports, higher landing fees[15] have been levied on aircraft that cannot meet the tougher certification standards; in some countries, aircraft without a noise certificate have been banned from operation[8,9] and, increasingly, airlines are being discouraged from buying older aircraft if a modern, quieter (but more expensive) alternative exists. All these moves have tended to mark down the value of second-hand aircraft, to the point where sometimes they have even been scrapped after being phased out of operation by the first owner. For example, the only de Havilland Tridents in service are operating in China; older B707 and DC8 aircraft have been excluded from operations in the United States and most of Europe unless they have been reengined or hushkitted to meet noise standards, and it is now likely that those types of aircraft that were given an environmental seal of approval only a decade ago will be phased out of operation by the early years of the next century.

Although much of the legislative activity in the 1970s was directed at the commercial jet fleet, other aircraft have also received a significant amount of attention. New designs of large propeller-powered aircraft, despite having to meet separate and less stringent rules for a number of years, are now required to comply with the requirements of Annex 16 Chapter 3 (i.e., the most stringent rule applied to jet aircraft). Smaller propeller-powered aircraft, usually in the business and general aviation sectors, are dealt with separately in a less expensive certification scheme, but helicopters are subject to the full process, under a rule similar to that for fixed-wing aircraft. The only aricraft to avoid the international legislative net are supersonic transports, of which the Concorde has been the only one to survive the rigors of service life. Since it is highly unlikely that a second generation of supersonic transports will materialise in the foreseeable future, although several nations have banned the sonic "boom", no provisions have been laid down to cover noise certification of this type of aircraft, other than in the United States.

All the above details are contained in Table 2.1. Major changes are not expected in the near future.

2.4   Operational controls

From the foregoing discussion, noise control can be summarised as a combination of actions taken to contain noise at its source within reasonable economic and technical limits (via certification), and then, as a separate exercise, to minimise impact once an aircraft is in service. In the latter context, over the past decade there has been a considerable upsurge in operational noise-control measures of the type listed in Table 2.2, at both national and local airport levels.

At the national level, large nations took positive action against the very noisy jets by first terminating their production and then phasing out of operation those already in the fleet.[8,9] That process, which started in the mid-1970s, has only just been completed. Although it has had some impact on airport noise levels by removing old B707, DC8, Caravelle and Trident aircraft, the fleet is still dominated by aircraft powered by low-bypass-ratio engines. The first illustration in this book highlighted the fact that the world fleet is dominated by two- and three-engined aircraft, mainly the B727 and 737 models and the DC9. The situation in the United States is typical of the world fleet, with over 2000 aircraft, or more than 50% of the commercial fleet, being powered by low-bypass-ratio engines that cannot satisfy the latest noise certification standards. In fact, most of them only just satisfy the earliest standards.

Table 2.1. *Summary of international aircraft noise legislation*

Aircraft type	International standards[a]	Significant deviations
Subsonic jets	Chapters 2 and 3, the latter being the tougher and now demanded of all new aircraft; Chapter 2, the original early-jet version of 1971	U.S. – the first version of FAR Part 36 was tougher than Chapter 2
Supersonic jets	None	U.S. has applied early subsonic rules; sonic boom prohibited
Heavy propeller-driven aircraft	Chapters 5 and 3, the latter now applicable to all new types	
Light propeller-driven aircraft	Chapters 6 and 10, the latter being the latest (and using ground-plane microphones)	FAR 36 version uses 1.2-m-high microphones
Helicopters	Chapter 8	not yet universally adopted

[a] Recommended by ICAO in Annex 16, for incorporation into member states' own legal framework

As a result, a second round of production and operational restrictions is now being considered. Europe, in the form of the twenty-two-member state European Civil Aviation Conference (ECAC) and the twelve-member-state EEC, have been vacillating on the topic for several years, closely watched by the United States. The moves of all nations have been hampered by oposition from nations affiliated with ICAO that rely on cheap, used older aircraft. ICAO does not intend to debate the subject until late 1989 and has asked its members to do nothing until then. Nevertheless, ECAC has agreed to a "nonaddition" rule, effective in 1990, and will probably undertake a phase-out programme after ICAO has let the protesters have their say. The current ECAC nonaddition plans centre on preventing airlines from buying more of the unwanted types by not allowing them to be registered and hence "added" to any individual European nation's register of aircraft. This at least puts a limit on the number of older aircraft owned by European airlines, but a phase-out programme is much more powerful in its impact on airport noise.

The United States formally joined the debate in 1985 when Congress instructed the FAA to report on ways of "accelerating fleet modernisation", in other words, arranging for the early demise of the JT8D engine. Although the FAA's reaction in 1986[29-30] was merely a cata-

Table 2.2. *Operational restrictions*

Operational restrictions or general limitations and penalties are applied in many areas of the world. These include the following:

"Nonaddition" rules: Here, a nation prohibits the addition of certain (older, noisier) aircraft types to its national register of aircraft.
Operational "bans": A nation deems that "no aircraft may take off or land unless . . . (it meets certain noise standards)".
"Nonproduction" rules: A manufacturing nation prohibits the production of certain (noisy) types.
Local rules: Aircraft are prevented from operating, unless they meet local airport rules. Benchmarks may be certification status, contour ("footprint") area or a locally imposed maximum noise level.
Night curfews: Noisier types are banned from using a particular airport at night.
Financial penalties: Airports will "grade" their landing fees according to noise nuisance, or a general noise "tax" will be applied to noisier types.
General restrictions: These include number of operating "slots" dictated by noise emission; preferential runway usage; take-off and landing timings; maximum noise levels to be observed ("policed" by monitoring systems).

Note: No regulations are applicable to military aircraft, but local operating guidelines may be used.

logue of options, with no stated policy preferences, it did talk separately to a small group comprising major airlines and airport operations, who came out with positive proposals in 1987.[31] Although these asked for some level of immunity to local airport rules in return,[52] they did point to the benefits of the type of proposals circulating in Europe, and suggested that the five million people around U.S. airports who experience 65 LDN or more would shrink to less than three-quarters of a million if the low-bypass-ratio engined fleet were retired. This, despite the projected 7% yearly increase in passengers carried, for it would result from the replacement aircraft being some 10 dB quieter and having a larger capacity.

One thing is clear: Any phase-out programme needs to be implemented well before the year 2010; otherwise it will happen of its own accord. We will have to await the outcome of the ongoing debate – the only positive move thus far being an intent by ECAC to start with a nonaddition rule late in 1990, and a similar EEC proposal.

At the local airport level, it is again the United States that shows the greatest activity. By 1987, more than 400 airports had imposed some form of noise-control restrictions. Their distribution is shown in Figure 2.4. For many years, the leaders in the field were the operators of Kennedy Airport in New York, the Port Authority of New York and New Jersey. They featured foremost in action taken against the high noise levels resulting from transatlantic operations in the late 1950s and early 1960s, and were violently opposed to the introduction of the Concorde (see Chapter 5). The visibility of the Port Authority has diminished in recent years, and the leaders have become the operators of John Wayne Airport,[10] in Orange County, California, who have to meet local state[11] requirements, and the proprietors of Washington National Airport,[12] which serves the nation's capital city, Washington, D.C. Both Washington's National and Dulles airports are owned by the FAA, Dulles having been built well away from the city to "take the noise from the people". National Airport noise is in part controlled by noise rules but also via limits on nonstop operational coverage. Although the radius of coverage permitted is now 1000 miles, aircraft coming from farther than 600 miles could not use the airport until quite recently. This kept the big jets out of National and forced them to use Dulles, but did not keep the noisier medium and small jets at bay, as is obvious to D.C. residents and visitors alike. Hence, it became necessary for the FAA to impose strict noise limits covering the night-time period, limits so strict they virtually acted as a curfew.

The night limits were never to exceed dBA levels on take-off and on approach, compliance being ascertained by conversion of the certif-

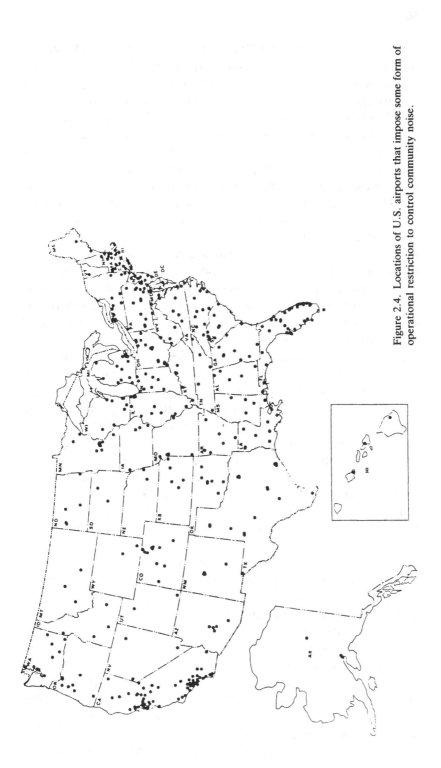

Figure 2.4. Locations of U.S. airports that impose some form of operational restriction to control community noise.

icated noise numbers from EPNdB to dBA. Several other airports, including John Wayne in Orange County, California, operate on the basis of noise "quotas". The system has worked successfully at Orange County for a number of years and other airports (e.g., Boston, Denver, Minneapolis) are now adopting similar policies. These allow the airlines to trade several "quiet" aircraft operations for each movement by a "noisy" type. Quiet and noisy are defined either on the basis of certification levels or by flight demonstration at the airport and the maintenance of good performance in the monitoring statistics. Under the latter conditions, every tenth of a decibel counts towards the profitability of an operation. With a never-to-exceed limit at the airport, rather than a judgment based on a previous and somewhat arbitrary certification test, it is possible to offload an aircraft to get inside the limit. This offload can either be passengers or fuel (i.e., range); conversely, a noise reduction will allow pay-load or range to increase. In fact, the exchange rate on modern aircraft at airports like Orange County can be as great as twenty passengers for every decibel! This is a very strong incentive for noise control.

However, local airport rules are extremely contentious and are a constant headache for those planning the route network, particularly if the airline has a mix of aircraft of different noise characteristics. Moreover, although it can be argued that airport proprietors ought to be able to impose their own operational noise limits, airlines that have bought the quietest aircraft they can find feel justified in seeking a high degree of operational freedom from local noise restrictions.[52] Equally, smaller airlines, with a large number of cheaper, older aircraft in their fleets, do not wish to be forced to purchase new aircraft ahead of any economic necessity, including the life of the aircraft structure. They may not be able to afford the capital investment necessary to purchase brand-new quiet aircraft, and it may be a number of years before quieter types are available second-hand. Most operators with a particular noise problem will, instead, try to operate their aircraft in the quietest mode possible, within the range stipulated by local air traffic control rules. Such operational techniques rely largely on one factor, keeping the power level of the engines as low as possible whilst close to a populated area. Unfortunately, such techniques often conflict with local air traffic control practices because they require a change in established operational patterns and are, therefore, mistakenly considered to be hazardous to safety. Nevertheless, for years, several airports have allowed or even demanded noise control via a reduction in engine power over populated areas, and the technique has long been permitted in the compliance-demonstration procedures of noise certification. The aircraft has to achieve a minimum safe altitude, dictated by airworthiness (safety) considerations and hence the number of engines fitted, but the pilot is then allowed to reduce power to a level at which safe flight can still be maintained if there is an engine failure. In practice, the altitude at which power reduction is selected and the degree of power reduction permitted are dictated by local conditions, particularly the distribution of the underlying population and local routing patterns. Heavy power reduction close to the airport will alleviate the highest noise levels, but it will restrict the aircraft climb gradient. Moreover, noise will increase downstream of the airport when high power is reselected for climb to cruise altitude. The principle of thrust management on take-off is shown in Figure 2.5, and the options for noise control on landing approach in Figure 2.6.

Noise control on approach is rather more difficult than take-off, for aircraft are generally tightly controlled by both air traffic procedures and the pilot, who is looking for a safe and stable aircraft configuration in the delicate manoeuvre prior to touchdown. This normally means that the aircraft is in an extremely "dirty" (maximum drag) configuration with full flap, landing wheels down and a fairly high thrust level. This enables the pilot to achieve maximum power very quickly, should it be needed in any emergency. But high thrust is high noise, and what are often referred to as "managed drag" configurations during approach are becoming more and more widely used. In this type of procedure, the final horizontal intercept to the instrument landing

Figure 2.5. Reduction of take-off noise by engine power cutback.

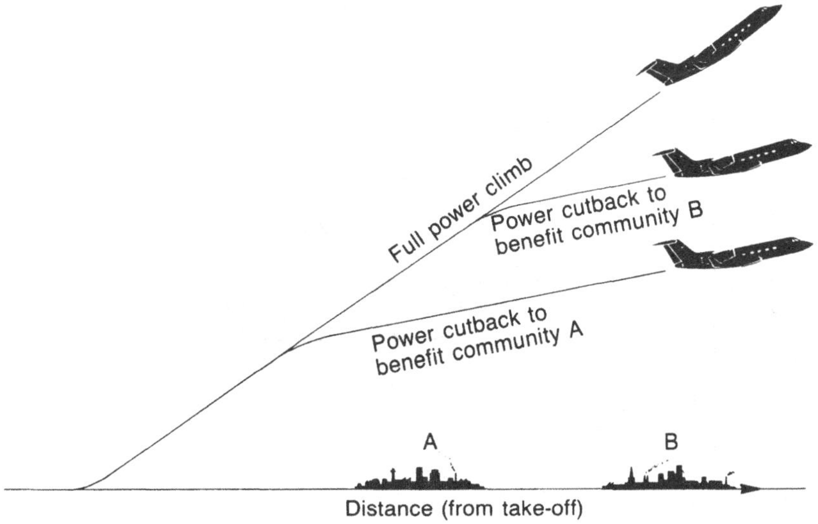

Distance (from take-off)

system (ILS) 3° glide slope (see profile A in Fig. 2.6) is replaced by a continuous descent at 3° (profile B), starting well out from the airport at high speed, with minimum drag and minimum power, and only ending up with the full-flaps, wheels-down, high-thrust configuration in the final phases of the approach. The rewards for using this procedure can be considerable. Figure 2.7 shows the range of noise reductions possible with a cross-section of older, noisier low-bypass-ratio-powered aircraft. Reductions of 5–10 EPNdB can be achieved some 4 km or more away from the airport but, unfortunately, the effect is lost when

Figure 2.6. Reduction of approach noise by aircraft speed/thrust/drag management.

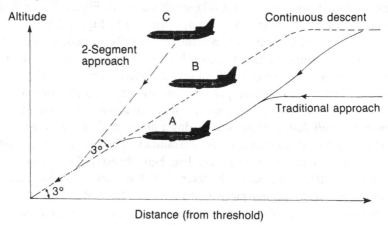

Figure 2.7. Noise reduction resulting from "managed" approach procedures.

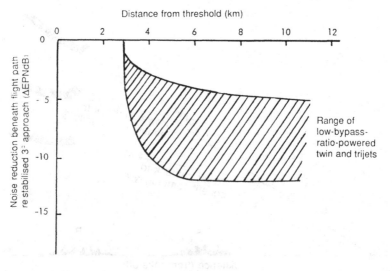

the aircraft is configured for the final stages of landing, just before the 2-km noise certification reference point. This control procedure is gaining acceptance by the airlines (which save fuel in the process) and by the airport owners.

An alternative has been proposed in the past, but it has proved to be unworkable.[13] This is the two-segment approach (profile C of Fig. 2.6), whereby the initial part of the descent procedure is carried out at a high angle (say, 6°) before joining the conventional 3° glide slope. Although tested and proved to be of considerable benefit from the noise standpoint, it relies on intercepting the 3° glide slope at a fairly low altitude, possibly only 150 m. This becomes a distinct operational hazard, the advantage of the conventional procedure being that if the 3° glide slope is "missed" in the level flight manoeuvre, there is no hazard, whereas if it is missed whilst performing a 6° descent, it could well terminate in a controlled crash!

Another procedure, practised at airports where the runways are long enough, is to displace the landing threshold so as to allow aircraft to overfly communities close to the end of the runway at a higher altitude. The gain in altitude over the community is around 10 m for every 200 m displacement achieved. More than twenty airports in the United States make use of this procedure.

Aside from the techniques described above, there are few others that will actually reduce the noise heard on the ground around an airport. Most others merely rely upon redistributing it to offload areas of noise saturation. Directional routing to avoid densely populated areas is a basic necessity at some airports, particularly those that have grown and encroached on developed areas. Alternatively, if the major communities are well away from the airport, then the best noise-control procedure is to climb to the highest altitude before a community is reached.

2.5 Land use planning

In this discussion, land-use planning[14] is a euphemism for keeping people and aircraft as far apart as possible. It can only be practised during the planning stage of a new airport or at a remote airport development. To exercise the option at a major airport is very difficult and expensive, but it has happened. In California, the Los Angeles Airport authority had to "purchase" a whole community and relocate people in order to build a third runway.

Several new airports have been built with land use in mind. On the outskirts of Paris, Charles de Gaulle Airport took the noise from the people. So did Dulles, near Washington, D.C., and Montreal's Mirabelle. But travellers do not like the extra hour or so that these airports impose on arriving, in-transit or departing passengers. London did not get a new airport on the eastern coastline, 40–50 miles from the city; Toronto did not get its out-of-town airport when Montreal's downtown Dorval was seen to be far more popular than Mirabelle and Washington's National Airport translates into only a ten-minute metro trip into town compared with the one-hour grind from Dulles. When this is doubled for an inward–outward visit to Washington, the penalty is considerable, both in time and money.

So, although laudable in its objective, land-use planning is not always popular. What is useful and acceptable is to limit the use of areas close to the runways to airport-associated activities, such as aircraft maintenance and airline storage, or light industrial activities. Normally, common sense can be trusted to determine what should be done with noise-impacted zones, but land values may tempt airport developers and local authorities alike. Indeed, in some cases, the highest rate of housing development can be found in the noisiest areas around an airport. The lobbyist for a new out-of-town airport may well have one eye on the commercial value of the land taken up by the "offensive" downtown airport. Also, although a good idea at first sight, a new airport usually develops its own business community, and then people want to live close to their jobs ... and the wheel comes full circle.

2.6   Other control measures

A number of other approaches have been tried from time to time, many of which are financially based. For example, some countries of Europe[15] and Asia levy increased landing charges on the operators of noisier aircraft when they use noise-sensitive airports; some airport neighbours may be compensated for diminished property values; others receive assistance in insulating their homes against sound.[16] Obviously, none of these actions actually makes aircraft any quieter, but they do help to suppress annoyance and are a popular ploy for administrations that do not wish to appear uncaring.

In the end, however, such measures are only pain killers, dulling the system until the long-term solution can be expedited – scrapping the older, noisier aircraft and building the best in noise control into the aircraft that replace them.

3

Aircraft noise sources

Noise, or unwanted sound, is generated whenever the passage of air over the aircraft structure or through its power-plants causes fluctuating pressure disturbances that propagate to an observer in the aircraft or on the ground below.[200-20] Since the flight condition[49] cannot be maintained unless these air- and gasflows are controlled efficiently, there are ample opportunities for sound to be produced. Fluctuating pressure disturbances result from inefficiencies in the total system and occur whenever there is a discontinuity in the airflow-handling process, particularly in the engines, where power generation involves large changes in pressure and temperature. This is not to say that the airframe itself is devoid of sound-producing opportunities, for it has a large surface area and, in the configuration that is adopted for take-off and approach, both the landing gear and high-lift devices (slats and flaps) create significant amounts of turbulence.

To the community beneath the aircraft, the self-generated noise from the airframe is normally significant only during the approach phase of operation, where the sources shown in Figure 3.1 can combine to exceed the level of each major noise source in the power-plant. For this reason, airframe noise has been thought of as the ultimate aircraft noise "barrier".[700] Perhaps we should briefly consider this and other airframe-associated sources before moving into the more complex subject of the aircraft power-plant.

3.1   Airframe noise

Since air flowing over the structure of the aircraft does so for distances of anywhere between a few centimetres (e.g., cockpit window surrounds) to many metres (e.g., the chord of the wing and the length of the fuselage), the scale of turbulence induced[242] varies considerably. Hence, broadband turbulence appears in the far-field* as a multifrequency noise signal, the degree of which varies throughout the spectrum according to the fundamental dimensions of the aircraft and the speed of the airflow over the structure. The sound so produced then has to propagate either through the atmosphere to the observer on the ground, or through the fuselage structure to the passenger in the cabin.

The passenger notices the presence of airframe-induced noise as the aircraft climbs and accelerates to cruise speed as a general build-up in the level of cabin noise. This build-up is proportional to about the fifth power of aircraft speed, although this is not fully appreciated since the reducing atmospheric pressure outside the passenger cabin (as aircraft altitude increases) offsets the effect by reducing the source level according to pressure squared.

The individual deployment of flaps, slats and the main landing gear are also quite recognisable from inside the aircraft. A substantial increase in cabin noise is experienced when the aircraft is prepared for the final landing approach. This results directly from the reaction of

Figure 3.1. Sources of airframe noise.

---

* That is, where the signal can be recognised as a sound-wave, usually over a wavelength from the source, and when the sound pressure and particle velocity are in phase.

the airframe to the turbulence induced by both the landing gear and the flaps.

Under normal flight conditions, with the undercarriage, slats and flaps retracted, the noise from the "clean" airframe is believed to be dominated by that generated by the wing.[703] Although the fuselage is a lifting body, most of the lift (and hence the drag,[704] induced turbulence and noise) results from the main wing and tailplane structures. Most of the noise is broadband in nature, although low-frequency tones have been noted, usually in association with cavities or discontinuities in the otherwise smooth surface. Landing-wheel bays[705] and modern high-lift flap systems are prime examples of such discontinuities, although a "clean" aerofoil can generate discrete tones.[706]

In practical terms, the only flight condition in which the aircraft configuration is truly "clean" occurs after the initial take-off and early climb, beyond the point where noise propagating to the observer on the ground is of concern. Conversely, as the aircraft climbs and accelerates, noise generated in the fuselage boundary layer becomes increasingly important in the passenger cabin. In a modern turbofan-powered aircraft, it is usually the main source of cabin noise, although a poorly designed cabin pressurisation/air-conditioning system can often be just as significant, as can engine noise that is either radiated through the air to the cabin wall or transmitted as vibration through the aircraft structure. To the observer on the ground, airframe noise reaches its greatest level during the landing approach, at which point the need to augment lift and deploy the landing gear for low-speed approach and touch-down increases the noise substantially.

Airframe noise research was particularly popular during the early days of the turbofan.[701-2] Between 1974 and 1978, a number of experiments were conducted in the United Kingdom under the auspices of the Royal Aircraft Establishment.[707-8] Many industrial organisations participated, and the results have been published. The results of one series of tests on a Lockheed L1011 TriStar[708] identified the main sources of airframe noise and investigated their dependence upon aircraft speed. Figure 3.2 is typical of the data acquired, and demonstrates a significant increase of up to 10 dB arising entirely from the deployment of the flaps and landing gear during the approach to land. Of these two sources, the landing gear produces the most intense noise, giving a spectrum-level increase of 5–10 dB and changing the directivity of the overall source to near-spherical (see Fig. 3.3). However, since these results were obtained on one particular airframe, the spectral character and variation of the noise signature with speed are a function of the design of that particular aircraft. The self-generated noise of a particular aircraft cannot be ascertained without an analysis of the mechanisms of generation associated with the specific airframe components. Such analyses have been conducted,[709-10] and the main conclusions can be summarised as follows:

> A clean airframe will produce a broadband source, distributed around a low-frequency peak, typically in the 200-Hz region, but it will vary somewhat according to the size of the aircraft and its speed. The most likely origin of the source is the main wing structure.

> Fairly strong tones may be observed from a clean airframe, associated with the vortex-shedding mechanism at the trailing edge of the wing.

> Theoretically, source intensity should vary according to the fifth or sixth power of aircraft speed. In practice, measurements are usually clouded by the noise of the aircraft engines, even at idle power, and velocity dependence appears to be lower (see Fig. 3.4).

> The broadband noise from the high-lift devices (trailing edge flaps and leading edge slats) and the landing gear is far more important than that of the "clean" airframe, increasing the

Figure 3.2. Comparison of measured airframe noise spectra from clean and gear-down aircraft configurations. (Crown Copyright – RAE)

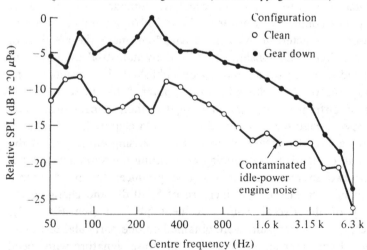

overall airframe noise level on the landing approach by about 10 dB.

Landing gear noise is the more important; it is broadly spherical in directivity (Fig. 3.3), has spectral characteristics slightly higher in frequency than those of the clean airframe (Fig. 3.2) and the level varies with about the fifth or sixth power of aircraft speed.

Figure 3.3. Longitudinal-polar noise radiation patterns (at 183 m) from Lockheed TriStar. (Crown Copyright – RAE)

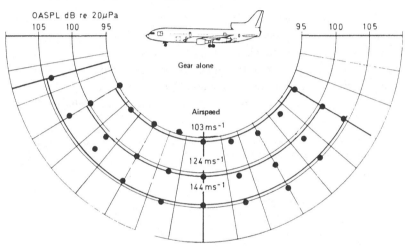

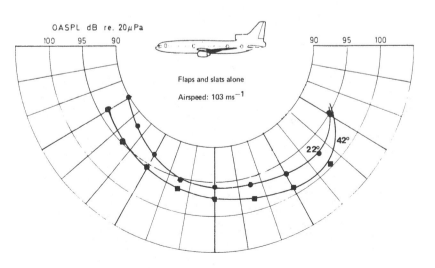

High-lift devices modify the spectral content of the "clean" airframe, tending to lower its characteristic frequency as a result of extending the chord of the wing and inducing larger-scale turbulence in the wing wake.

3.2   Sonic boom

Sonic boom is another form of airframe noise, in that it is a direct result of the operation of an aircraft in the atmosphere, albeit at supersonic speeds. Military aircraft are the major producers of sonic booms; among civil aircraft, only the few operational Concordes produce booms.

The sonic boom is the result of an observer sensing the passage of the pressure or shock wave that any finite body causes when it travels through the atmosphere at supersonic speeds.[720] In the case of aircraft, it consists of a minute but sharp rise in pressure, followed by a steady drop over a fraction of a second, with an immediate return to normal atmospheric pressure. Because of the general shape of the

Figure 3.4. Increase in noise caused by the deployment of gear and flaps, as a function of airspeed. (Crown Copyright – RAE)

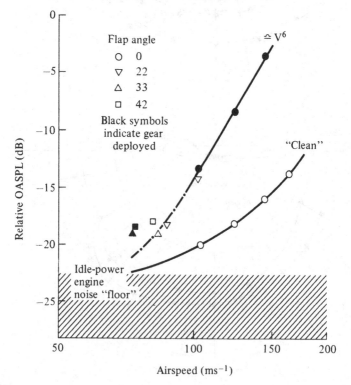

pressure–time history, the boom is often described as an "N" wave. If sustained as a cyclic and repeatable pressure disturbance, it would be over 100 dB but, because of its short duration, the impulse sounds like a short sharp crack of thunder. To be really noticeable, the instantaneous "overpressure" has to be of the order of one-twentieth of 1% of normal atmospheric pressure, or about 50 Pa, although lower overpressures do cause a reaction, particularly if the ambient noise is low and the boom particularly "sharp". Physically, the boom is analogous to the bow wave created by a ship. It is generated by the compacting effect on the pressure wave that the aircraft produces as it disturbs the still atmosphere. All aircraft produce a pressure wave, but when they are flying at subsonic speeds the wave moves away in all directions at the speed of sound and is too weak to be audible. It is the engine one hears when an aircraft is cruising overhead.

When the speed of the aircraft is supersonic, the pressure waves cannot get away ahead of the aircraft (Fig. 3.5), as their natural speed

Figure 3.5. Development of a sonic boom.

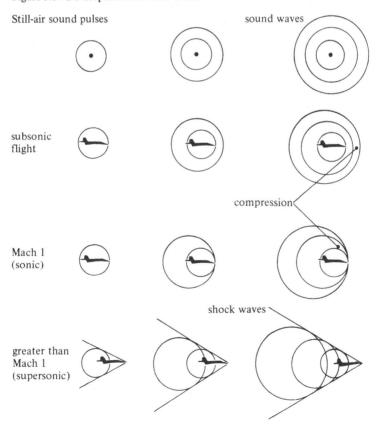

is slower than that of the aircraft. Slower, in this context, means just over 1200 km/hr at sea level and about 10% less at normal cruising altitude. Because they cannot get away, the pressure disturbances coalesce and lag behind the aeroplane, which is in effect travelling at the apex of a conical shock wave.[721-2] The main shock wave is generated by the extreme nose of the aeroplane, but ancillary shocks are generated by all the major fuselage discontinuities. Hence, the detailed pressure–time history of the boom is quite complex, as it consists of a series of minor perturbations within the general N shape of the overall pressure–time history. Figure 3.6 shows a typical wave form close to the aeroplane compared to that received between 100 and 300 aircraft lengths away (i.e., at typical cruising altitude). The short multiple N-wave pattern close to the aircraft becomes diffuse at larger distances and resolves into an almost single N-wave over a period of 0.1–0.2 sec.[723] Data from the Concorde[735] and other supersonic aircraft suggest[48] that the initial boom overpressure normally ranges from 25 to 150 Pa, and that public reaction varies from "negligible" to "extreme" over this range[729-35] (see Fig. 3.7). At the upper end of the range there is also concern about the long-term structural integrity of buildings that are likely to be subjected to repeated sonic booms.[733-4]

Figure 3.6. Degeneration of the sonic boom pressure–time waveform with distance from the aircraft.

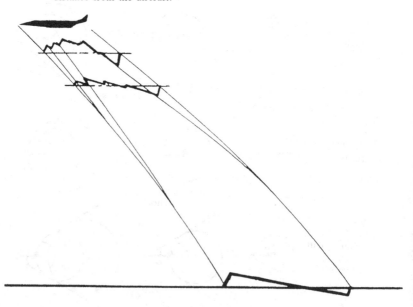

The severity of the boom can be controlled,[830-3] in part, by appro-
priately altering the aerodynamic design of the aircraft fuselage and
wing sections, and by controlling aircraft speed, altitude and incidence
to the flight path.[724-7] However, the boom cannot be avoided entirely,
and thus many countries do not allow a supersonic transport to pass
over their territories.[5,728] Even a modest boom can startle human and
animal life in remote areas when it emanates from a clear blue sky –
and possibly more than once. The rise time of the overpressure is an
important factor, since this determines whether the boom is a sharp
crack or a distant rumble.

Depending upon atmospheric conditions at the time, the boom can
propagate to an observer on the ground in one of three ways (Fig.
3.8). Firstly, it may move directly through the atmosphere from the

Figure 3.7. Measured sonic boom overpressures.

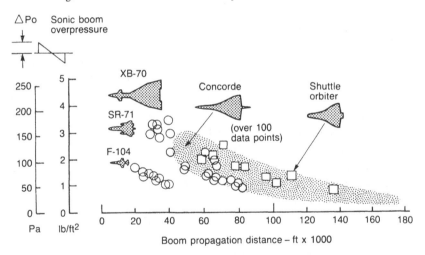

Figure 3.8. Three modes of propagation of the sonic boom: (A) direct, (B)
refracted secondary, (C) reflected and refracted secondary.

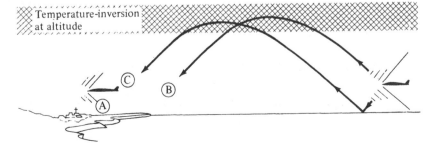

aircraft to the observer, and then possibly via one of two secondary routes[732] – the upward wave may be refracted back from regions of temperature inversion at altitude, and the downward wave may be reflected, firstly upwards from the ground and then it too may be refracted back from the upper atmosphere. In the operation of Concorde between Europe and the United States,[32] careful attention was paid to avoiding the direct mode of propagation by limiting departure speeds to the subsonic régime over land, and decelerating from supersonic to subsonic speeds well away from the destination. These precautions did not, however, prevent the secondary modes of propagation from disturbing the peace from time to time.[732] In fact, many complaints were not known to be connected with the sonic boom until it was found that they resulted from the secondary modes of propagation.[732] These effects can only be avoided through reliable prediction of upper atmospheric conditions and careful control of aircraft attitude and speed.

Although several countries have prohibited civil supersonic flights over their territories, if a new supersonic transport is to become commercially viable in the future it is almost certain that these regulations will have to be relaxed somewhat. The worldwide network of routes take aircraft over many land masses and inhabited islands and, although it is extremely doubtful that supersonic aircraft could be designed to produce no boom at all, an "acceptable" overpressure of about 25–50 Pa might be achieved. This would require careful design of the aircraft and a strictly controlled operational pattern that did not produce any sudden variations in either speed or aircraft attitude.[22] Cruise altitude might well have to be increased considerably.

Like all aircraft, the supersonic transport cannot fly continuously at high-altitude steady cruise conditions. Speed, altitude and angle of attack to the flight path vary during the accelerating climb to cruise conditions, just as they do during the deceleration and let-down from cruise altitude prior to landing approach. It is under the climb and descent conditions that flow patterns around the aircraft are likely to depart most substantially from the ideal, and that the sonic boom is likely to be of greater intensity than has been designed for in the cruise condition.

The details of the boom can be predicted from design geometry, which has shown, in principle, that it is possible to moderate the overpressure at cruise conditions.[830–3] For example, the angle of attack, or attitude of the aircraft, and its distance from the observer (altitude) are probably the two most important factors in determining the magnitude of the boom overpressure. An aircraft's angle of attack is directly related to the amount of lift it must produce to sustain its own weight; a heavier aircraft requires a larger amount of lift and, with a wing of a given size, a larger angle of attack. This generates a strong shock wave, with substantial overpressure arriving at the ground.

As is evident from Figure 3.7, the altitude of the aeroplane influences the intensity because of the distance the shock wave has to travel before reaching the ground. Weather conditions also affect the way in which the boom propagates. Wind speed and direction and air temperature and pressure all influence the direction of travel, and the strength of the shock wave and air turbulence near the ground may cause considerable distortion of the regular N wave. In addition, hills and valleys, or even tall buildings, can cause multiple reflections of the shock waves and affect perceived boom intensity.

Another important effect is "lateral dispersion", which is similar to the distribution of engine-generated noise close to the airport (see Chapter 7). By definition, an observer to the side of the aircraft flight path has a greater distance between himself and the aircraft than someone beneath it, and therefore the shock pattern has further to travel and will be relatively weaker. Figure 3.7 includes data from Concorde development flying taken to the side of the flight path and at very large distances. However, because of wind and temperature gradients in the atmosphere, sometimes the lateral observer may not hear the shock at all. At shallow angles of incidence to the ground, the shock waves are often refracted back into the upper atmosphere and never reach the ground.

Nobody can predict the impact of repeated booms, despite the special experiments set up for this purpose.[729] Therefore, perhaps it is fortunate that only a handful of SSTs are in service and that a new design is unlikely to appear for many years.

3.3   The propeller

Aircraft propulsion systems are without doubt the major sources of aircraft noise. Propulsion systems come in many forms: the jet, the turbofan, piston- and turbine-driven propellers, the helicopter rotor and the developing "open-rotor" or advanced, high-cruise-speed propeller.

This latest version of the propeller brings the wheel full circle, as the propeller played an important part in the earliest powered flying machines. From the simple and, by modern standards, inefficient spar of wood to the modern composite open rotor or "propfan", the sources of noise have been the same, albeit generated in different proportions and at different intensities. The laws of physics remain unchanged. Moreover, these sources are fundamental to noise generation in the complex geometry of the multibladed turbomachinery of the gas turbine engine, on which are superimposed other internally generated sources and, of course, the noise from the external jet mixing with the atmosphere. Each should be examined because, nowadays, they are important factors in determining the detailed design of the aircraft power-plant. Let us start with the propeller.

Prior to World War II, the world's modest commercial airline fleet was composed entirely of piston-engine propeller-driven aircraft. It was not until the appearance of the novel Rolls-Royce Dart–engined Vickers Viscount and the revolutionary de Havilland Ghost–powered Comet in 1952 that the gas turbine had any part to play. Because the world fleet was modest in size, interest in propeller noise was only superficial or academic; in fact, during the war, the louder the exhaust or propeller of a friendly aircraft the greater the comfort to the nation at war. Nevertheless, some of the mechanisms that generated propeller noise were known as far back as the 1920s,[600] and, before the advent of the civil gas turbine engine, the theoretical and experimental re-

Figure 3.9. A counterrotating open rotor propulsion system for high cruise speeds propulsion system.

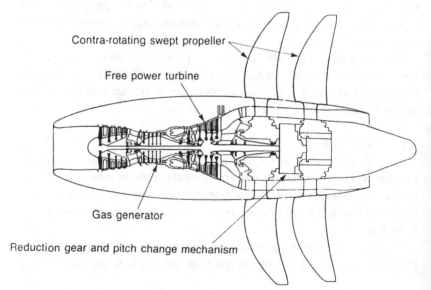

Contra-rotating swept propeller

Free power turbine

Gas generator

Reduction gear and pitch change mechanism

search in Europe and the United States had laid the foundations for predicting the components of discrete tone noise. The importance of broadband noise had also been registered.

With the subsequent development of the gas turbine engine for jet- and turbofan-propulsion systems, studies of propeller noise lost some of their appeal. In fact, for many years propeller-powered aircraft were regarded as "quiet" and were given far more freedom of operation than (noisy) jet-engined aircraft. It was not until the introduction of noise certification, around 1970, and its subsequent application to propeller aircraft, that research in this area picked up again. The recent renewal of interest in propeller noise has been driven largely by propulsion concepts like that shown in Figure 3.9, which use advanced swept multiblade propellers for flight speeds as high as those attained by the commercial jet fleet. This work, and the preceding work on straight-bladed propellers,[604–16] has led to a greater understanding of the mechanisms of generation. This knowledge has been applied in the development of quieter conventional propellers.

3.3.1 Noise sources

Like the majority of propulsion-system noise, propeller noise falls into the tonal (or discrete frequency) and broadband categories.

Tones are produced as a result of the regular cyclic motion of the propeller blade in the atmosphere with respect to a stationary observer, and by interactions with adjacent structures. The frequency spectrum comprises the tone at the fundamental blade-passing order and its higher harmonics, with the harmonic decay rate being a function of the blade form and its operating duty.[607–9] The spectrum in Figure 3.10 is typical of a conventional propeller with an average harmonic decay rate of 5–10 dB. In general, the near-field signal has a basically sinusoidal pattern at low bladespeeds, which becomes "sharpened" as a result of incipient shock formation as the relative velocity of the blade tip (helical tipspeed) approaches the speed of sound.

The broadband component arises from random fluctuations in pressure over a wide range of frequencies, and is associated with turbulent flow in the inlet stream, in boundary layers and wakes behind the blades.[604–6] Fundamentally, the dimensions of the blades and the velocity of the flow over them will determine the spectral shape and, although it is independent of bladespeed, the peak level often appears at around ten times the basic frequency of the blade-passing tone. This is because blade dimensions and numbers are not entirely unrelated in a propeller designed for a given power level.[608] However, compared to the energy in the discrete frequency components, broadband noise is often insignificant. The sources of tone noise are, therefore, very important[621] and, in the main, arise as a result of the following:

> Air-volume displacement effects, as each blade disturbs the air to produce a regular pulse with respect to the stationary observer. This is frequently referred to as "thickness" noise,[627-8] which becomes most important at high bladespeeds.

> Changes in blade-section motion relative to the observer as the steadily loaded propeller rotates, generally referred to as "loading" noise.[606] This source tends to dominate at low bladespeed.

> Localised effects when the bladespeed approaches or exceeds the speed of sound.[623]

> Periodic unsteady loadings resulting from pressure-field and nonuniform inflow effects, for example, when the rotor disc is set at incidence to the direction of free-stream flow,[650-1] or senses "upwash" close to the leading edge of the wing.

> Other interaction effects, notably blade-wake impingement on other (fixed or rotating) structures,[652-3] such as the wing or the second blade row in a counterrotating design.

Figure 3.10. Typical propeller noise spectrum, showing propeller tone harmonic content and decay rate.

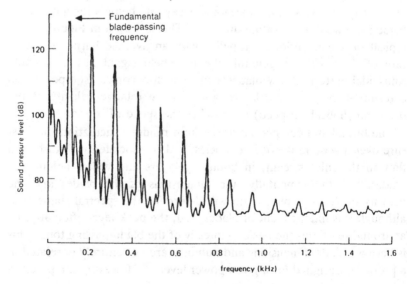

Designs that feature two rows of blades[640-1] produce powerful effects, according to bladespeed, blade geometry and row-to-row spacing.

The loading and thickness noise sources were first recognised over fifty years ago.[600] Both sources appear at the same frequencies, blade passing and its harmonics, but thickness noise assumes increasing relevance with increasing rotor speed and harmonic number. As regards loading noise, the lift forces (which are steady in a coordinate system) produce cyclic pressure changes at a fixed point in space due to blade rotation. Thickness noise simply results from the regular parting of the air.

When the Mach number of the blade relative to the local airflow (helical tip M_n) approaches unity, the localised effects in the fluid around the blade can be a source of intense noise. Early supersonic-tipspeed propellers were notoriously noisy[602] but, because the local noise sources tended to be associated with a highly inefficient aerodynamic régime, they failed to be developed. That is why conventional straight-bladed propellers are not used for high-cruise-speed civil applications, and why the modern open rotor, or propfan, is becoming important.[620-32] These designs have very thin blade sections and are swept back at the tip, so that the component of velocity normal to the leading edge of the blade is subsonic, even when the helical tipspeed is supersonic. This design criterion minimises tip effects and thus suggests that the inherent advantages of the propeller – in terms of its propulsive efficiency, overall weight and cost – might well be extended to flight speeds hitherto the province of the jet. Demonstrator versions of the modern open rotor started flying in 1987, and they could be in commercial service by the mid-1990s, that is, providing the cabin-noise problems historically associated with propeller-powered aircraft can be solved, along with the mechanical/safety and aerodynamic issues.

Periodically induced unsteady loadings generate sound in much the same way as the steady loading. Here, however, the loading is not fixed in any frame of reference, a fact that is important in counter-rotating two-stage propellers, typified by some old straight-bladed designs[640] and the latest swept blades of the propfan concept. In these designs, interaction tones resulting from unsteady loadings can appear at both upstream and downstream row blade-passing frequencies (and their harmonics) as the rows sense the presence of one another. This can result either from pressure-field interactions or, in the case of the downstream row, wake and tip-vortex effects.

The same is true of the single rotor running at incidence to the flow[650-1] or close to a fixed structure[652-3] (e.g., the wing) but, here, the fixed structure has a blade-passing frequency of zero! Nevertheless, the unsteady loading sources still exist, appearing at multiples of true blade-passing frequency. In fact, wing upwash-induced incidence effects can also cause asymmetry of the noise radiated to each side of the propeller – often the difference between port and starboard being of the order of 5 dB (see Chapter 7).

3.3.2   Control of propeller noise

It has been recognised for many years that, for single-row propellers with subsonic helical tipspeeds, noise is reduced when tipspeed is reduced or, by virtue of the reduced loading, when blade numbers are increased.[601] However, in the latter case, this does not automatically mean that the noise perceived by the observer is reduced, since increasing frequency increases annoyance at frequencies below 3–4 Hz. Nevertheless, the trade-off is often in favour of less annoyance, and that is why modern aircraft are tending to use an increasing number of blades. For example, the latest BAe ATP has six-bladed propellers in place of the four-bladed design of the old BAe 748.

Moreover, because of the powerful effect that blade-tip section and profile have when tipspeeds are high, considerable attention is paid to the shape of the blade tip. Equally, the distribution of the loading, both span and chordwise, is scrutinised to avoid areas where local flow changes could result in (avoidable) high levels of sound generation. None of this is simple, and the extension to the two-row machine only emphasises the problems.

Although each row can be considered independently, and best-design criteria applied to each, the interaction between the rows is of considerable importance. To date, the standard noise-control techniques are as follows:

> *Increasing row-to-row spacing:* This minimises both potential field and blade-wake interactions.
>
> *Increasing blade number:* This can push some harmonic tone-energy "off the scale", that is, above the audible range or into the range above 3–4 Hz, where it is less important psychoacoustically and more rapidly absorbed by the atmosphere.
>
> *Differentially selecting blade number:* This can influence the annoyance created by avoiding high energy in the same frequency bands, where blade numbers are equal.

Reducing tipspeed: This is done to reduce blade Mach numbers, as in the single-row case.

3.4   The "jet" engine

The word "jet" was originally a colloquial term for an aircraft powered by propellerless gas turbine engines of the type that first entered commercial service in the Comet in the 1950s. It now also covers aircraft with modern turbofan engines since these, too, have no visible propeller and rely upon the energy contained in the flow that is exhausted from a nozzle (or nozzles) located at the rear of the engine.

In the period of forty years since the Comet first flew, the jet engine has gone through significant changes in design philosophy and operating cycle. The original jet was a single-shaft, single air-and-gas-flow system with one (circular) nozzle at the exit. It relied upon compressing a comparatively small mass of air, which, after combustion and extraction of energy to drive the compressor system, was released into the atmosphere as a hot jet at high velocity. The operating cycle of the jet engine is exactly the same as that in an internal combustion engine, except that the process is continuous rather than cyclic as the pistons in a reciprocating engine compress a fixed amount of air each time the shaft rotates. This is illustrated in Figure 3.11.

In a typical 50-kN-thrust engine, around 40 kg airflow per second is drawn into the compression system, mixed with fuel and then burned. Downstream of the combustion system, the turbine extracts sufficient energy to drive the compressor, which is on the same single shaft, before the hot gas is expelled through the nozzle at around 600 m per second. Net propulsive thrust is simply the residual energy in the jet after overcoming the opposing drag forces on the engine nacelle.

The acceleration of a small mass of air to high velocity is not the most efficient aerodynamic process, but it was the only process possible with the materials available to the inventors of the pioneer breed of jet engine. High jet velocity is also an essential feature of the propulsive cycle for flight at very high speeds (e.g., in supersonic passenger transports and high-speed military aircraft). It was not until better materials were developed and the advent of turbine-cooling technology allowed blades of increased efficiency to be made in comparatively small sizes, that the "bypass" cycle saw an improvement in propulsive efficiency. The bypass cycle requires about 50% more air per unit of thrust, and therefore reduces the mean exhaust velocity. Shown in Figure 3.12(b), the bypass engine divides the larger entry airflow into two almost equal streams, utilising only one to supply a smaller but

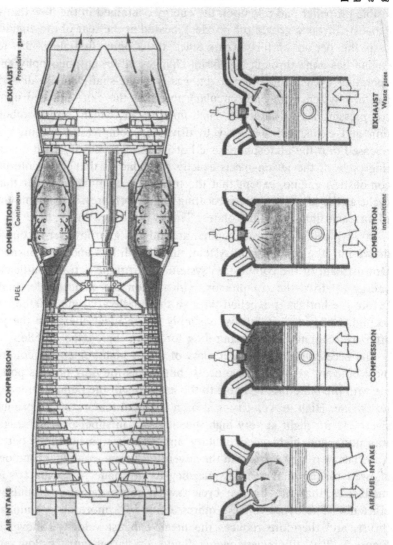

Figure 3.11. Comparison between the working cycle of a piston engine and that of a gas-turbine.

Figure 3.12. Aero gas turbine design cycles and their applications.

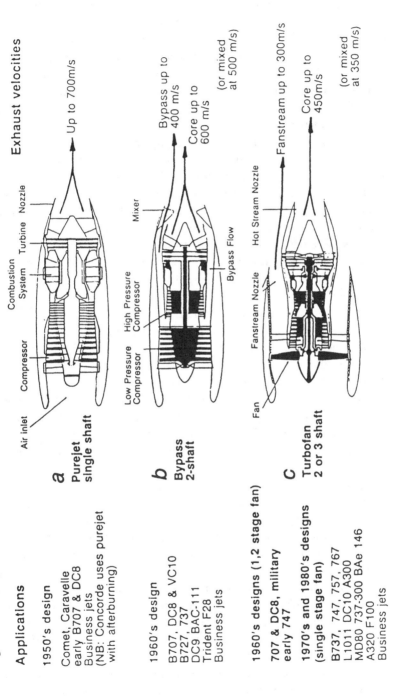

Applications

Exhaust velocities

1950's design

Comet, Caravelle
early B707 & DC8
Business jets
(NB: Concorde uses purejet
with afterburning)

Air inlet

Compressor

Combustion
System Turbine Nozzle

a

**PureJet
single shaft**

Up to 700m/s

1960's design

B707, DC8 & VC10
B727, 737
DC9 BAC-111
Trident F28
Business jets

Low Pressure
Compressor

High Pressure
Compressor

Mixer

Bypass Flow

b

**Bypass
2-shaft**

Bypass up to
400 m/s

Core up to
600 m/s

(or mixed
at 500 m/s)

1960's designs (1,2 stage fan)

707 & DC8, military
early 747

**1970's and 1980's designs
(single stage fan)**

B737, 747, 757, 767
L1011 DC10 A300
MD80 737-300 BAe 146
A320 F100
Business jets

Fan

Fanstream Nozzle

Hot Stream Nozzle

c

**Turbofan
2 or 3 shaft**

Fanstream up to 300m/s

Core up to
450m/s

(or mixed
at 350 m/s)

more efficient gas-producing engine core, which then has the same fairly high exhaust velocity as in the pure jet. The remainder of the entry airflow is held at a lower pressure, and bypasses the primary core flow in a separate duct to be expelled through a separate nozzle at about two-thirds of the hot gas velocity. Alternatively, the hot and cold flows can be mixed within the engine carcase, to further improve propulsive efficiency, before being exhausted through a single nozzle at a lower (mean) velocity.

The modern turbofan, shown in Figure 3.12(c), takes yet another step forward in efficiency by using even higher levels of bypass flow via a large fan, driven by even more efficient air-cooled turbine units operating at higher gas pressures and temperatures. In fact, it is a salutary thought that on the latest turbofans the compressor exit air temperature (or combustion system entry temperature) is now the same as the combustion exit temperature was on the earliest engines! Moreover, the combustion exit temperature is now so high that the gasflow is turning a rotor assembly that would melt if it were not for the cooler air circulating within each blade.

This technology allows three to four times as much air to be involved per unit of thrust as was used in the pure jet, with only 20–30% being used in the gas-producing process upstream of the turbines. The large and, nowadays, single-stage fan provides modest compression (less than 2 : 1) and hence fan-jet exhaust velocities of only some 300 m/sec. The core compression ratio is much higher than on earlier engines, often well over 20 : 1, but there is also much higher energy extraction in the turbine to drive the bigger fan and higher pressure compressors. As a result, the hot jet velocity is lower than on any earlier type of engine, at only 400–500 m/sec.

Again, in some turbofan engines, both the hot and cold flows are mixed ahead of a single propulsive nozzle by using a high-efficiency mechanical mixing arrangement, or by allowing the two streams to mix naturally in a comparatively long exhaust duct. This process reduces both the peak nozzle exhaust velocity and also the associated mixing losses.

Since all sources of aerodynamic noise are a function of velocity, as bypass ratio rises and jet velocity falls, so the external jet mixing noise falls also. Conversely, the power handling in the turbomachine rises with increasing bypass ratio and so do the noise levels generated by the fan, intermediate pressure (IP) compressor and turbine system. This change of source emphasis alters not only the character of the noise but also the directivity pattern of the sound field. Low- and high-bypass ratio radiation patterns can be compared diagrammatically, as illustrated as in Figure 3.13, and the resultant changes in absolute noise level can be verified by reference to measured noise data. In Figure 3.14, published "lateral" or "sideline" noise certification levels have been notionally corrected to the same engine take-off power and then plotted on a base scale of bypass ratio, which is the ratio of the cold fan duct massflow to the hot core flow. The externally generated

Figure 3.13. Diagrammatic comparison of noise radiation patterns from low- and high-bypass-ratio engines.

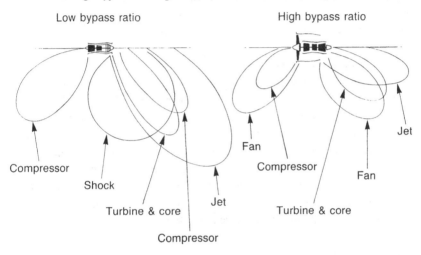

Figure 3.14. Variation of source noise with engine bypass ratio.

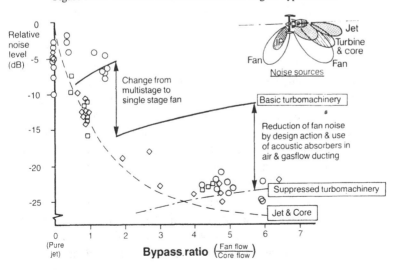

jet noise "floor" provides the lower noise limit at any given engine design cycle. Perhaps surprisingly, it becomes apparent that, without the beneficial effects of noise control in the basic design of the turbomachine and use of acoustic absorbers in the air- and gasflow ducting, the modern turbofan would still be almost as noisy as the early pure jet and bypass engines. Moreover, the generalised case shown in Figure 3.14 is true only for the maximum power case, for the noise sources in a jet engine vary with both power level and engine design concept. At lower powers, the importance of jet exhaust noise is reduced and, in the case of high-bypass-ratio turbofan engines, the fan, compressor and turbine noise sources dominate. Therefore, it is necessary to work through the engine and consider each of the main sources in turn.

3.5   Fan and compressor noise

Like all rotating machines, a fan or compressor system will exhibit sound that has both tonal and broadband characteristics. It is the tonal element that has attracted most attention over the years, since piercing high-frequency discrete tones often characterised the air-compression systems of early jets, particularly during the approach phase of operations, and both low- and high-frequency tones have been evident during all phases of operation of the modern turbofan. Since tonal and broadband sound are generated by quite different mechanisms, they need to be discussed separately.[421]

3.5.1   Broadband noise

The broadband source results from the propagation of sound produced near the surface of the blade as a result of the pressure fluctuations associated with turbulent flow in the vicinity.[422–46] Turbulence is induced whenever there is a flow of air or gas over a solid surface, or there is a discontinuity between gaseous flows. The important sources of turbulence in the fan system are the boundary layer in the outer wall of the inlet duct and the wakes shed from each stage of blading. In addition, although turbulence in the atmospheric inflow is not normally significant in the generation of broadband noise, the presence of obstructions in the inlet duct can be. As a result, most engines are now designed with the absolute minimum of mechanical features that might generate additional turbulence.

In the first stage of the compressor – or the fan in a turbofan engine – broadband noise is generated in the movement of the tip of the rotating blade within the turbulent boundary layer close to the wall of the inlet duct. Here turbulence levels are high and, by definition, the tipspeed of the blade is at a maximum. Turbulence in the wakes shed by fan blades (Fig. 3.15) is also an important source of random noise, particularly for fan blades with large surface areas, and plays a significant role in the generation of broadband noise in the downstream stages of the turbomachine. In a modern single-stage fan, these are the outlet stators in the bypass duct and the multistage core compressor. In a multistage fan, there will be significantly greater "activity" in the subsequent rotating and stationary stages, and broadband noise levels will be higher.

The importance of turbulence in the mainstream flow in which an aerofoil is operating was first demonstrated several years ago in a simple laboratory experiment.[402] An aerofoil was suspended firstly in the laminar core flow from a model jet exhaust and then moved into the turbulent mixing region. Broadband noise-level changes of the order of 15 dB were observed. Subsequent experiments on full-scale engine compressors and independent theoretical analysis all showed a clear relationship between turbulence intensity, air/surface velocity and noise. Early published experimental data on aeroengine fans and compressors first quantified the situation,[405-6] confirming the theoretical fifth-power variation of acoustic energy with velocity of the airflow local to the surface. It also showed that the mean velocity of flow through the compressor dictated the split of radiated energy from the air inlet (at the front of the compressor) and from the exhaust nozzle (at the rear of the engine). The work also indicated that results from a single-stage fan exhibited much lower levels at a given tip velocity, due to the lower level of free-stream turbulence. Most recent

Figure 3.15. Generation of broadband noise.

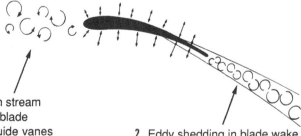

1. Turbulence in stream approaching blade (e.g., from guide vanes or upstream stages) causes random fluctuations in lift pressure over blade surface and generation of random or broadband noise

2. Eddy shedding in blade wake reacts back on blade to cause random lift-pressure fluctuations over blade surface and random noise generation

experimental work has concentrated on the accurate definition of sound levels, and a description of the frequency spectra radiated both forwards and rearwards from a single-stage fan at defined operating conditions. For the high bypass-ration turbofan type of blading, a clear relationship has been established between absolute sound level and aerodynamic efficiency. Previously, a variation in level had been noted at a given airflow velocity when the angle of inflow incidence was altered, with variations in absolute level of around 1–2 dB per degree of incidence change being common[851] (see Fig. 3.16). The more recent work has explained this variation and expressed the absolute level recorded in the far-field in terms related to the energy dissipated in the creation of the turbulent wake behind the blade.

3.5.2   Discrete tones

Discrete tones are generated when there is an interactive effect between airflow perturbations in the path of a rotating blade row, or "stage". Most frequently, these interactions involve pressure-field or turbulent-wake disturbances, but they can be associated with interactions with natural atmospheric turbulence or result simply from the propagation of the unique pressure pattern associated with each blade, particularly where supersonic flows are involved. Because the whole subject of tone generation and propagation is extremely complex,[400–41] we can do no more than give it broad coverage in this

Figure 3.16. Effect of a 4° change in blade incidence on generation of broad-band fan noise.

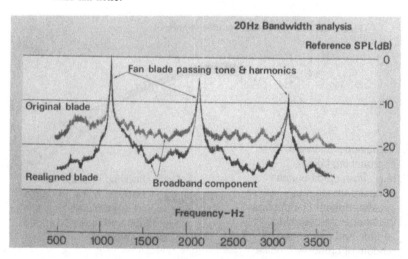

section. Extensive references are provided for those wishing to pursue the subject in detail.

The simplest mechanism by which a discrete tone is generated is the propagation to the far-field of the almost identical pressure fields associated with each blade on a single rotating stage.[430] The situation is analogous to the unducted propeller, previously discussed, but under subsonic tipspeed conditions that prevail at low engine-power settings, the energy involved is low and often masked by other effects. At high power, particularly during the take-off phase of aircraft operation out of an airport, where the modern fan is running with supersonic tipspeed, the effect can be marked. Because the airflow is supersonic relative to the region of the blade tip, there is a stand-off shock wave ahead of each blade, analogous to the sonic boom from an aircraft travelling at supersonic speeds. Providing the shocks from each blade are all identical, they will propagate to the far-field observer to give a discrete tone at blade-passing frequency, and its higher-frequency harmonics. However, for this to happen, not only do the blades have to be identical mechanically, but their aeroelastic properties have to be such that they all behave in the same way under the extreme loads of high-speed rotation. Clearly, this is unlikely to occur in practice for, even if it were possible to create identical physical shapes at the manufacturing stage, it is highly unlikely that their aeroelastic properties would be identical at design conditions, where each blade suffers a 50-tonne centrifugal load in delivering a tonne of thrust. Moreover, in service, erosion and other "foreign object damage" do not allow blades to remain identical for the life of the fan. Hence, the shock strengths and directions normally vary from blade to blade, as illustrated in an exaggerated manner in Figure 3.17.

Figure 3.18, which is not exaggerated, shows the pressure pattern ahead of a sector of ten blades of a thirty-three-bladed fan, obtained during several revolutions. Although each pressure pattern is quite repeatable, no two are identical. Hence, under normal circumstances, each blade shock will be heard to a greater or lesser degree in the far-field, as tones from single-blade order (shaft speed) through to blade-passing order, and as their harmonics above blade-passing frequency. As Figure 3.19 illustrates, the spectrum is a veritable "forest" of discrete tones. However, at the large distances normally involved when one is listening to engines at maximum power, the higher-frequency harmonics are absorbed naturally by the atmosphere, and the lower shaft-order tones associated with the stronger shocks tend to predominate. The overall effect is similar to the sound of a high-revving two-stroke motor-cycle or a sawmill, and is often referred to as "buzzsaw" noise.

Buzzsaw noise is also quite audible in the cabin of an aircraft under take-off conditions, in that part which is ahead of the fan. In this case, the high-frequency harmonics are absorbed, not by the atmosphere, but by the fuselage, its acoustic treatment and the internal cabin trim, whereas the low frequencies propagate through the cabin wall and can

Figure 3.17. Diagrammatic representation of mechanism of buzzsaw noise generation.

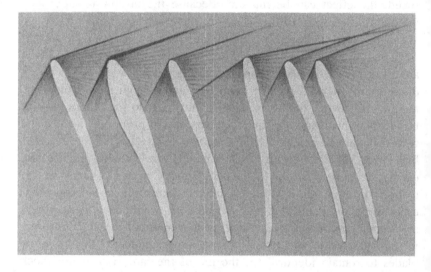

Figure 3.18. Pressure–time signature ahead of fan operating with supersonic tipspeed.

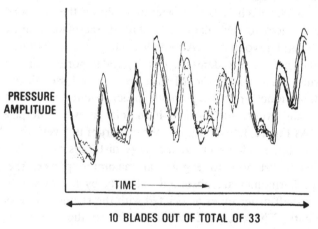

frequently be heard throughout the aircraft climb. The effect diminishes in intensity as pressure decreases with increasing altitude and, by the time the aircraft has reached cruise altitude, it is very difficult to detect buzzsaw tones.

At subsonic conditions, when no shock pattern exists, it is the cyclic pressure-field and wake interactions between rotating and stationary stages that are the source of discrete tones.[400-1] If the stages are close together, there is generally an intense pressure-field interaction and a very strong tone is generated (Fig. 3.20(b)). To avoid this, and possible failure modes arising from blade stresses, the rotating and stationary rows are normally spaced in such a way that the intensity of the pressure-field interaction is below that of the wake interaction. The intensity of the discrete tones then becomes a function of the strength of the wake behind the upstream stage, whether this is the rotating or stationary stage (Fig. 3.20(c)). The frequency at which tones are observed is a function of the number of blades on the rotating stage and the speed of its rotation. The pressure patterns so produced, their propagation direction and the multiplicity of the wave-front interference effects give rise to a lobular far-field distribution pattern which, in three dimensions, is analogous to the surface of a pineapple or an extremely deformed golf-ball.

The elements important in defining the details of the propagation pattern are the number of blades on the stationary and rotating rows, their relative speeds and the absolute flow velocities in the interacting cascades. Like broadband noise, intensity is a function of the magni-

Figure 3.19. Typical buzzsaw noise spectrum.

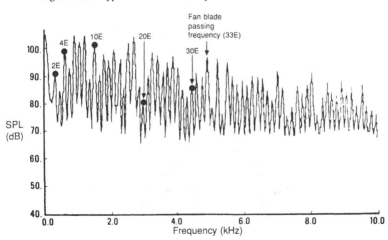

tude of the pressure disturbances, including those that are caused by
any steady distortion pattern in the intake flow. Such distortion might
be produced by an eccentric or otherwise nonsymmetrical intake design, or by the inclusion of aerodynamic features upstream of the fan.
Propagation effects are more concerned with the geometry of the
situation, which can be extremely complex. Therefore, let us consider
some less complex cases, involving very few rotors and stators, of the
order of 10% of the number that would constitute a full-scale fan.

Consider the simple case of a four-bladed rotor (Fig. 3.21). When it
is spinning adjacent to a stator row having only one blade (a most

Figure 3.20. Mechanisms of generation of discrete tones in a compressor
stage. (After Ref. 211, by permission)

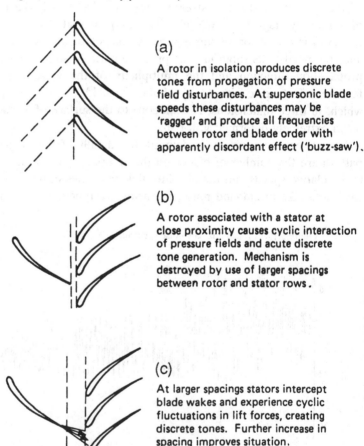

(a)

A rotor in isolation produces discrete
tones from propagation of pressure
field disturbances. At supersonic blade
speeds these disturbances may be
'ragged' and produce all frequencies
between rotor and blade order with
apparently discordant effect ('buzz-saw').

(b)

A rotor associated with a stator at
close proximity causes cyclic interaction
of pressure fields and acute discrete
tone generation. Mechanism is
destroyed by use of larger spacings
between rotor and stator rows.

(c)

At larger spacings stators intercept
blade wakes and experience cyclic
fluctuations in lift forces, creating
discrete tones. Further increase in
spacing improves situation.

Figure 3.21. Interaction tone generation; rotor blade numbers an exact multiple of stator vane numbers. (From Ref. 211, by permission)

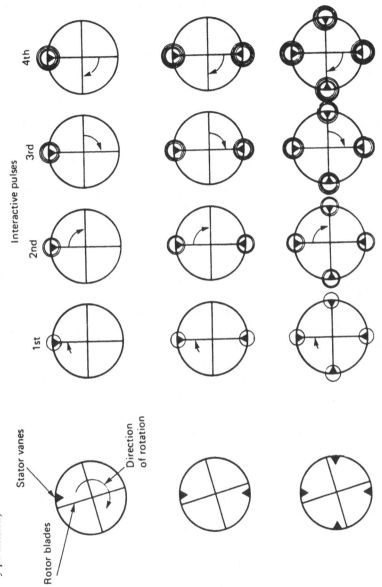

unlikely situation), there will be an interaction "pulse" every quarter-revolution, as each rotor blade (R) passes the stator vane (S). This will produce a tone at blade-passing frequency originating at the location of the stator. With two equispaced stator blades, the process will occur simultaneously in the region of each stator, and with four equispaced stators it will occur simultaneously four times during each quarter-revolution. This multisource pattern complicates the situation, for it also gives rise to amplifications and cancellations in the far-field due to interference effects.

If the number of stators is an inexact ratio of the number of rotor blades, which is the usual situation in a compressor, the phenomenon becomes more complex. With one stator less than the number of rotors (Fig. 3.22(a)), each quarter-revolution of the four-bladed rotor induces three interaction pulses, equispaced in time but at the position of each successive stator, and occurring in space and time in the direction of rotation. Hence, one full rotation will produce not only a regular interaction at each stator location four times per revolution, but also a superimposed pattern of events that is taking place more rapidly (at four times rotor speed) and occurring in the same direction of rotation as the blading. Even if the rotor is moving subsonically at the tip, the rate of occurrence of the threefold pattern of events can be supersonic. Interestingly, when there is one more stator blade than the number of rotor blades (Fig. 3.22(b)), the pattern of events still occurs at four times the rotor speed, but in this case it moves counter to the direction of rotation of the blading. This can be significant in controlling the propagation and radiation of fan tones through the upstream blade row.

This pattern of events can be expressed quite simply. If the number of rotor blades is B and that of stator vanes V, a pattern occurring at a harmonic number n times the blade-passing frequency can be formed in m circumferential wavelengths, where

$$m = nB \pm kV$$

(k being any integer), and will rotate at

$$nB/m = nB/(nB \pm kV)$$

times the rotor speed. In the case of the four-bladed rotor and three-bladed stator above, $n = 1$ and for $k = 1$ the speed of the pattern becomes

$$4/(4 - 3),$$

or four times the rotor speed.

Figure 3.22. Interaction tone generation; inexact multiples of rotor blades and stator vanes.

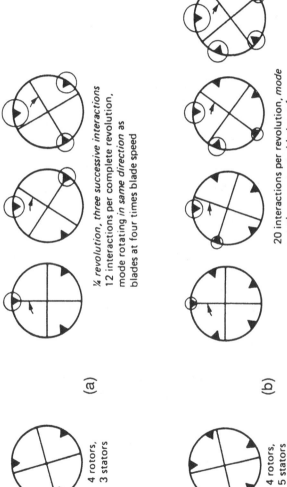

(a)

¼ revolution, three successive interactions
12 interactions per complete revolution,
mode rotating *in same direction* as
blades at four times blade speed

4 rotors,
3 stators

(b)

20 interactions per revolution, *mode
rotating counter to blades* at four
times blade speed

4 rotors,
5 stators

Most actual fans and compressors have about twenty to fifty blades and even more stators. The analysis is accordingly more complex, and has to include consideration of duct-propagation criteria. Indeed, in an unconstrained situation, although most tones do propagate beyond the confines of the duct, some do not. There is a critical ratio of stator/ rotor blade numbers beyond which certain dominant modes can be induced to rotate such that they decay within the duct, and hence do not propagate to a farfield observer. Known as the "cut-off" condition, it can be simply expressed such that the number of stator vanes (V) has to exceed

$$1.1 \ (1 + M_n) \ nB,$$

where M_n is the local Mach number. Hence, in order to cut off the fundamental ($n = 1$) tone in a fan operating with a tipspeed close to the speed of sound, there need to be just over twice as many stators as rotors; and to cut off the second harmonic ($n = 2$) tone, over four times more stators than rotors are needed.

It will be appreciated that, with a multiplicity of rotating and stationary blade rows, the noise source is extremely complex. The frequency spectrum of a total "front-end" engine assembly contains discrete tones from both the fan and early stages of the core-engine compressor, both at their fundamental blade-passing frequencies and higher harmonics, and at intermediate frequencies. The tones at intermediate frequencies, often referred to as "sum and difference tones", result from the interaction between persistent wakes from early stages of blading with subsequent downstream stages, or effects resulting from acoustic waves interacting with other rotating blade rows. In the typical engine assembly shown in Figure 3.23, tones propagating down the fan duct will normally be a function only of the interaction between the fan blade and its outlet guide vanes (OGVs), whereas the tones propagating from the inlet are far more numerous and complex in their origin. Wakes from the fan blade will interact not only with the fan OGVs, but with core-engine stator vanes (S1, S2, S3, etc.). There will also be basic interactions between the wakes from the core-engine stators and rotors (S1-R1-S2, S2-R2-R3; S3-R3-S4, etc.) and, where the fan blade is rotating on a different shaft from that of the core-engine compressor, interactions between the fan-blade wakes and the rotating stages of the compressor (F-R1, R2, R3, etc.). Fortunately, it is difficult for tones generated well into the core-engine compressor to propagate upstream against both the flow and the physical "barrier" of the upstream blade rows, which resemble a series of venetian blinds to the wavefront propagating upstream.[410-11] The blind may be "open" or "shut", depending on the wavefront angle and whether it is propagating with or counter to the direction of rotation of the blading. This is an effect that now forms part of the turbomachinery noise-control armoury, along with cut-off and rotor–stator blade-row separation. Even so, a spectrum similar to that shown in Figure 3.24 is not unusual. Fan tones (F) and compressor rotor tones (R) interact to produce a significant number of sum tones.

3.5.3   Interaction tones resulting from atmospheric turbulence

For many years, the importance of the interaction between rotating machinery and natural atmospheric turbulence went unrecognised. Throughout the 1960s and into the 1970s, many fan tests were conducted with little regard for the quality of the air entering the test unit. It was not until fans designed for fundamental tone cut-off were seen to benefit under flight conditions, but not under static conditions, that the reasons were examined in detail.[460-7] One particular experiment demonstrated the problem beyond any reasonable doubt and laid the foundations for the understanding that led to the technical requirement for inlet "flow conditioning" on all static test arrangements.[468-9]

The experiment involved the measurement of the noise signal from an RB211 turbofan installed in a specially adapted VC10 flying test-

Figure 3.23. The turbofan engine front end: a typical fan and compressor configuration. (From Ref. 211, by permission)

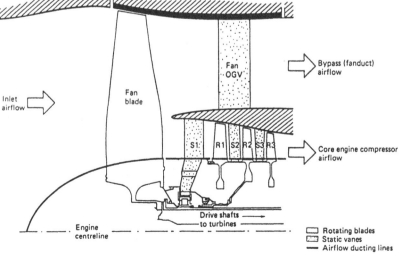

bed, shown in Figure 3.25. In order to gain as much information as possible on any changes that took place between the static and moving environment and with increasing flight speed, numerous microphones were installed in the inlet duct and in the fan duct of the power-plant and also extensively in the flap-track housings on the trailing edge of the wing. Since the engine was mounted on the rear of the aircraft fuselage, it was possible to carry out an extensive survey of the noise propagated forward through the inlet, as typified in Figure 3.26, which presents data from one of the inlet duct microphones. Here, the upper half shows 30-sec steady-state recordings under both static and flight conditions (at about 200 kn) for the same modest engine power setting. It is clear that the fluctuating nature of the fundamental tone under static conditions is completely eliminated and the tone becomes virtually steady under forward speed conditions. There is also a considerable drop in absolute level, on average over 5 dB and, periodically as much as 15 dB. But, since the fundamental interaction tone between the fan and its outlet guide vanes on this engine was designed to be cut off, it can be argued that no tone should be present at all. In fact, the fan "bypass" tone was cut off, for subsequent work showed the

Figure 3.24. Spectrum containing sum tones (fan and first-stage compressor interactions: F, fan; R1, R2, compressor rotor number; 1R2, first harmonic of R2).

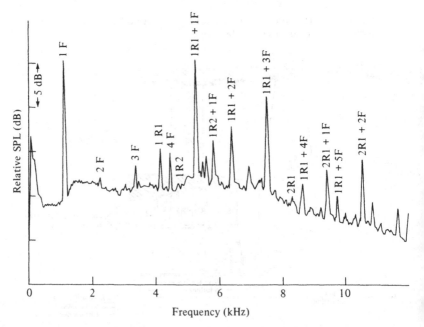

Figure 3.25. A Vickers VC 10 modified to act as an RB 211 Flying Test-Bed.

Figure 3.26. Variation of RB211 fundamental tone level during static and flight tests of VC 10 FTB.

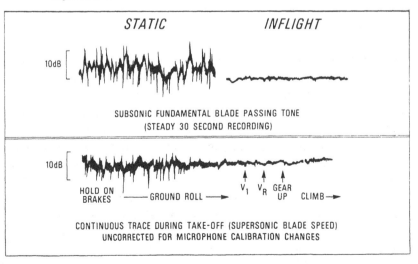

observed tone to be the result of an interaction between the fan and the core-engine compressor. Had this interaction not been present, the elimination of the interaction tone between the fan and its bypass-outlet guide vane would have resulted in a 20-dB drop to the level of the broadband background.

The lower half of the illustration shows how the tone character varied with increasing forward speed. From the "hold" on the brakes and through the early part of the ground roll (to a speed of about 60 kn) the tone was still fluctuating, but beyond that point it became progressively steadier. The fact that the level appears to rise with time in the illustration should not be confused with the statement that the tone was actually reduced, for it merely reflects the lack of correction for changing pressure conditions in the vicinity of the microphone cartridge, as both aircraft speed and altitude were increased.

It was this experiment that led to close consideration of the changing environment between the static and flight conditions, and the realisation that the unsteady nature of the fundamental tone was a direct result of an interaction between the fan and atmospheric turbulence, and had nothing to do with the design of the fan and its outlet guide-vane system. In short, under static conditions, natural atmospheric turbulence and turbulence created by structures close to the engine are drawn in from a nearly spherical capture area, and various sectors of the fan disc see random changes in pressure/velocity as large turbulent eddies are digested. Under flight conditions, where turbulence levels are generally lower anyhow, the effect of a stream-tube "capture" of the inflow provides for more uniform ingestion and a much reduced pressure/velocity variation across the fan face.

Once this situation was understood, it was recognised that the meaning of much of the accumulated static test data, both on engines and on compressor rigs, had been obscured by turbulence interaction effects. Consequently, a concerted effort was made to design inflow control systems that would more nearly reflect the atmospheric environment of an engine under its normal flight operation conditions. Inlet flow-control screens of the type illustrated in Figure 3.27 became a feature of aerodynamic model-scale test facilities, and it is now common practice to undertake fan-noise studies only where inflow control is available.

The control devices are all very similar, being little more than wind-tunnel flow smoothing screens "wrapped" into a spherical shape. This is the clever part. The problem of making a fully self-supporting device measuring over 6 m in diameter for full-scale engine testing, such as that in Figure 3.28, which must not introduce its own sources of turbulence and which could double its weight when soaking wet, is not to be dismissed too lightly! Nevertheless, all major manufacturers now find it essential to equip their engine-noise test facilities with such devices, which have to be carefully calibrated to take into account any acoustic transmission losses. Fortunately, these are normally no more than a decibel.

3.6   Turbine noise

Turbine noise, like fan noise, is a combination of tonal and broadband signals.[492] However, since the spacing between adjacent rotor and stator rows is generally much less than in a fan, it is tone noise that dominates the situation.

The geometry of the turbine for the three-shaft Rolls-Royce RB211 is illustrated in Figure 3.29. Behind the combustion chamber, a series of nozzle guide vanes (NGVs) direct the high-energy combustion gases at supersonic speeds onto the high-pressure (HP) rotor blades. Here, sufficient power is extracted to drive the high-pressure compressor system, which is on the same shaft as the turbine. The flow is then directed via another set of nozzle guide vanes onto the intermediate-pressure (IP) turbine blades, mounted on a rotor that is connected by a separate shaft to the intermediate-pressure compressor. Downstream at this stage the flow is then directed onto the blades of a three-stage

Figure 3.27. Design of an inlet flow-control screen for a compressor rig.

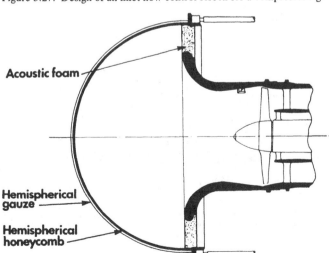

Figure 3.28. A 6-m-diameter inlet flow control screen on an open-air test-bed.

Figure 3.29. Typical turbine assembly layout – three-shaft engine.

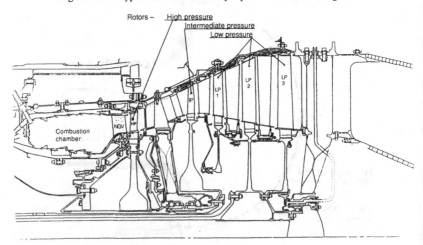

low-pressure (LP) turbine mounted on yet another shaft, where sufficient energy is extracted to provide the power for the single-stage fan at the front of the engine, which provides 80% of the engine thrust.

High levels of energy handling take place within the turbine, often in excess of 0.1 MW, and there is ample opportunity for the generation of interaction tones between the five (or more) rotating and stationary stages of the turbomachine.[494] Turbine stages tend to be fairly closely spaced together and, since the generative mechanisms of tonal noise within the turbine are identical to those within the fan and compressor system,[493] potential field interactions often dominate the lesser blade-wake interaction mechanism.

The multiplicity of interaction tones is well illustrated in Figure 3.30, taken from a two-shaft, four-stage low-bypass-ratio engine unit. On this particular turbine, the dominant fundamental tone from the final low-pressure stage was clearly audible during the approach phase of operation, and the beneficial effect of increasing rotor–stator spacing (as with compressors) was demonstrated to be a practical solution to the problem.

The main differences between turbine noise and compressor or fan noise lie in the following elements:

Since the nozzle guide vanes behind the combustion chamber run "choked", it is not possible for the acoustic signal to propagate upstream and, therefore, all the energy is radiated from the exhaust nozzle.

In this radiation process, the wave fronts have to propagate through the shear layer of the jet mixing with the atmosphere and therefore they undergo refraction.[490-1] As a result, dominant turbine noise radiates only at angles of about 110° to 130° to the intake axis (50° to 70° to the exhaust axis), as illustrated in Figure 3.31.

In propagating through the turbulent shear layer, the otherwise sharply defined tonal nature of the spectrum becomes more diffuse and the individual tones become broadened, or "haystacked".[495] Frequently, they are not recognisable as anything but humps in the spectrum and may be misinterpreted as broadband in origin. The thickness of the jet shear layer compared with the acoustic wavelength is important in dictating whether haystacking takes place, for fan tones or turbine tones

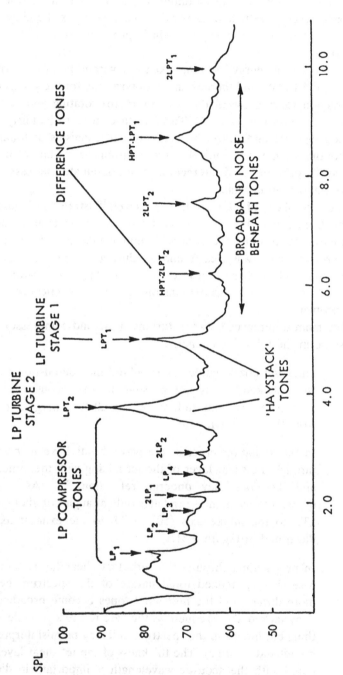

Figure 3.30. Spectral character of turbine noise.

Figure 3.31. Schematic radiation patterns of internal engine sources, showing refraction of sources radiating from the exhaust nozzles. (From Ref. 211, by permission)

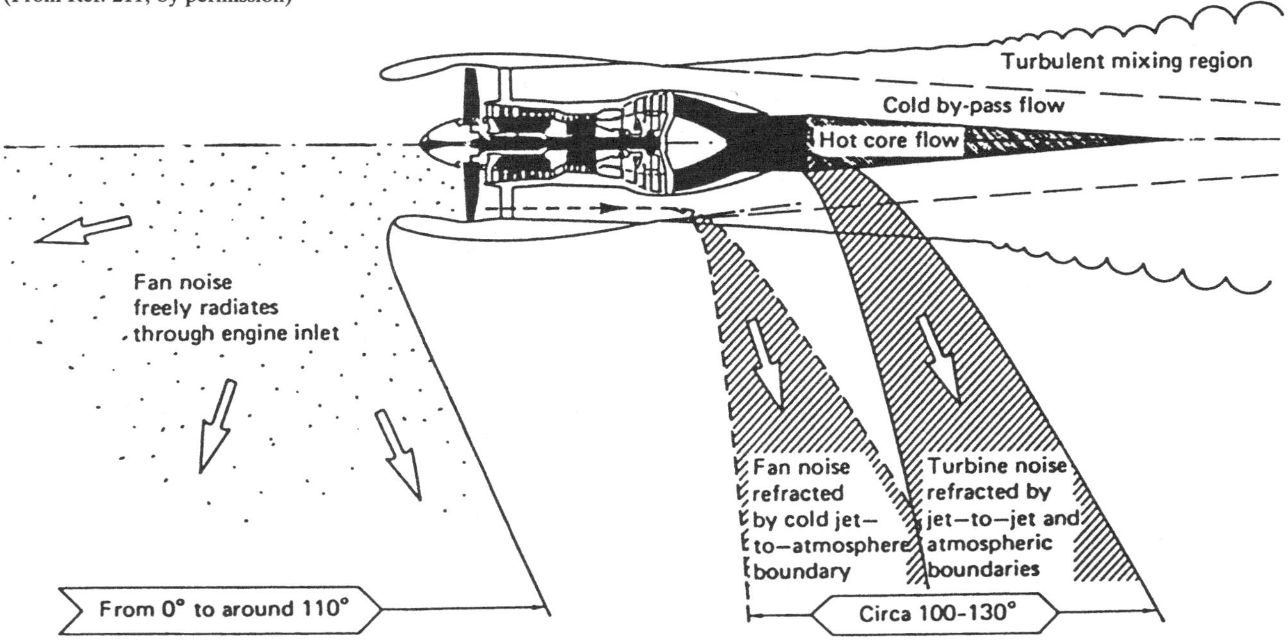

Turbulent mixing region

Cold by-pass flow

Hot core flow

Turbine noise refracted by jet—to—jet and atmospheric boundaries

Circa 100-130°

Fan noise refracted by cold jet—to—atmospheric boundary

Fan noise freely radiates through engine inlet

From 0° to around 110°

generated by (cold) aerodynamic models rarely suffer hay-stacking. This is well illustrated by the comparison of the spectra acquired from both a model turbine and its full-scale engine counterpart, shown in Figure 3.32.

The exhaust geometry can also have a marked effect on the "sharpness" of the turbine tones, depending upon the degree of internal mixing of the fan and core flows, and the way in which it is induced.

Because of the close proximity of each rotating and stationary stage – and, indeed, of each subsequent stationary and rotating stage – there is ample opportunity for the appearance of "sum and difference" tones, as in the fan/IP compressor interface area. These are also well displayed in Figure 3.32.

Figure 3.32. Turbine noise spectra measured on a cold aerodynamic model rig and a full-scale engine. (From Ref. 211, by permission)

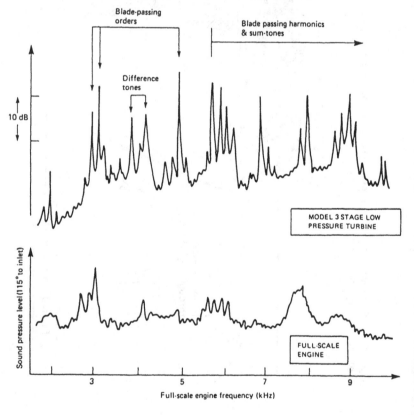

There is also the same opportunity in the turbine as in the compressor to take benefit from acoustic cut-off. In fact, it is generally easier than with the fan or compressor because of the combination of the much lower tipspeeds and a higher speed of sound in the hot medium. Cut-off can usually be achieved with far less than twice as many stator blades as rotor blades, and it is now common practice to take advantage of this phenomenon in the design of the modern turbine.

The combustion system upstream of the turbine can have an influence on the generation of tones. No combustion system produces perfect circumferential uniformity of either temperature or velocity, and there can be very powerful interactive effects between the nonuniform flow field out of the nozzle guide vanes and the early stages of the turbine.[497]

3.7 Jet noise

The expression "jet noise" has become part of the international language, and is usually taken to mean the noise of jet-powered aircraft. However, strictly speaking, it covers only those sources associated with the mixing process between the exhaust flow of the engine and the atmosphere, and those components associated with the shock system in an inefficiently expanded jet of supercritical velocity.

The shock system is vividly displayed in Figure 3.33, where the luminosity variations that accompany the pressure changes in the reheated, incompletely burned, exhaust flow serve to make the process visible. The normal mixing process is less visible, although the use of oil injection to create smoke in the jet can give some feel for the less ordered and extremely turbulent nature of the process, as shown in Figure 3.34. The jet-velocity decay patterns associated with the ideally expanded and "shocked" jets are sketched in Figure 3.35.

Over half the world's fleet of commercial aircraft are still powered by engines with a take-off noise signature that is characterised by one or more of the various jet sources. All too often, areas near major airports suffer the intrusion of the "crackling"[310] and "tearing" sound created by high-velocity jet exhaust flows and, because the shock-associated component radiates as strongly in the forwards direction as it does after the aircraft has passed by, the noise persists for long periods.

The origins and spectral characteristics of the jet mixing and shock-associated noise sources are illustrated in Figure 3.36. Of course, a spectrum taken from an engine measurement will include not only

Figure 3.33. Visible presence of shock cells in an underexpanded supercritical jet exhaust.

Figure 3.34. Smoke visualisation of the jet mixing process.

Figure 3.35. Velocity decay rates in ideally expanded and underexpanded "shocked" jets.

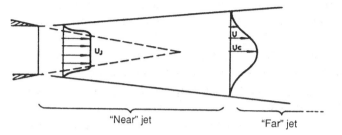

"Near" jet   "Far" jet

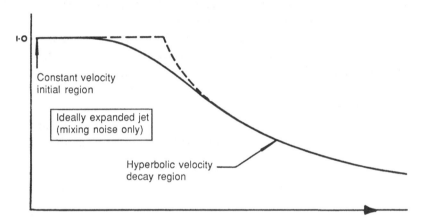

1·0

Constant velocity
initial region

Ideally expanded jet
(mixing noise only)

Hyperbolic velocity
decay region

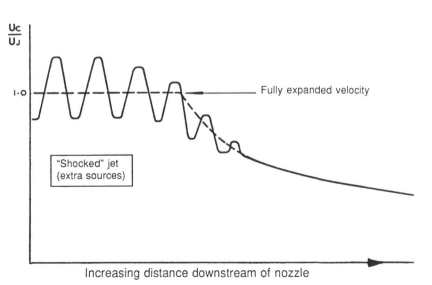

$\dfrac{U_c}{U_J}$

1·0   Fully expanded velocity

"Shocked" jet
(extra sources)

Increasing distance downstream of nozzle

these components but also all the noise from the core engine, which comprises the turbine and combustion systems, and so on. However, since none of these "exhaust" noises result from the expansion of the jet as it mixed with the atmosphere, they are dealt with elsewhere in this chapter.

For jets operating in the sub critical régime (i.e., with an exhaust velocity less than the local speed of sound), the mixing noise is the only component to be generated. The shape of the frequency spectrum is a reflection of the fact that the eddies that comprise the turbulent mixing process[240-1] vary considerably, increasing in size progressively downstream of the nozzle and decaying in intensity as the velocity falls and the mixing becomes complete.[316-20] In the case of the super-critical jet, shock-associated noise appears as a superimposed secondary source of a largely broadband nature. However, so-called screech tones have been observed in engine exhaust tests and are often a common feature of experimental work on cold model jets.

3.7.1   Shock-associated noise

The mechanisms that generate both screech and broadband shock-associated noise[300-9] have to do with the expansion shocks. The

Figure 3.36. Origin of shock and mixing noise components of jet noise spectrum.

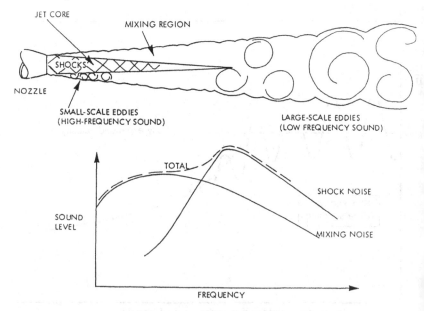

tones are believed to originate through a feedback mechanism between the shocks and the nozzle lip, whereas the broadband component has been shown to arise as a result of turbulent eddies interacting with the shock structure. Although the mechanisms are fairly well understood, it has proved difficult to define the parameters that control source amplitude. Screech has been found to be common in cold model jet experiments, but it has seldom been important enough on engines to be a cause for concern. There has been some evidence of tailplane structural fatigue[304] where the engines are mounted on the rear of the fuselage, but the problem has normally been handled by fitting a jet "suppressor" nozzle to eliminate the shock structure. Alternatively, it is possible to use a convergent–divergent nozzle designed to expand the jet at nearer ideal conditions at the most critical pressure ratio in the aircraft's operational envelope. Con–di nozzles cannot be designed to cover a range of pressure ratios, unless they are fully variable.

From the standpoint of community noise, it is only the broadband shock noise that is of concern. Although it is generated more or less omnidirectionally, because of the intense level of mixing noise to the rear of the engine, the impact of shock-associated noise has usually been most marked forwards of the aircraft during the take-off operation. In the forward quadrant, shock-associated noise can quite easily exceed mixing noise by 15 dB, and be more or less of an equal amplitude in the rearward quadrant. Fortunately, like the discrete component, it is also amenable to suppression by disturbing the ordered shock structure with a jet suppressor or by eliminating the expansion cells by use of a convergent–divergent nozzle.

Unlike the discrete component, the intensity of the broadband shock-associated noise is reasonably predictable. It has been shown that intensity varies only as a function of pressure ratio, being independent of jet temperature (and hence velocity), and it is also essentially independent of observation angle.[307] The spectral character, already illustrated in Figure 3.36, can be described loosely as broadband, but it does tend to have a distinct spectral peak, as a result of interference effects between the radiation pattern from the many shock cells present in the jet. The total process was described several years ago,[306] and the main observations were expanded to form the basis of the first published prediction procedure. This was subsequently refined after further analysis and comparison with measured aircraft data, and it now forms the basis of the most complete prediction method available.[947]

3.7.2 Jet mixing noise

Historically, this is the source of engine exhaust noise that has received most attention and about which there is a veritable library of theoretically and experimentally based dissertations.[300-71] It is the source that gave rise to the aircraft and airport noise problem of the 1960s, and that has been least amenable to reduction by engineering means, both mechanical[820-3] and aerodynamic.[824-5]

About thirty-five years ago, the early theory[201] showed clearly that the fluctuating shear stress in the mixing process behind the nozzle would generate broadband noise. This theory states that intensity should vary according to the eighth power of the exhaust velocity (V). In practice, most experimental work has indicated significant departures from the V^8 relationship both at high and low velocities. At low velocities, any observed departure is a function of sources other than jet-mixing noise and high-quality experimental work reveals that the V^8 relationship holds. At high velocities, the departure can be explained in terms of the speed at which the source eddies are convected, being comparable to and even in excess of the speed at which the acoustic waves propagate to the far-field.[323] Figure 3.37 illustrates the variation of exhaust noise with jet velocity, indicating that as jet velocity becomes much greater than the speed of sound, source intensity becomes governed by kinetic effects and tends towards V^3.

These data also indicate the increasing departure of V^8 of hot model jets and whole-engines as jet velocity falls below the local speed of sound. Although much of the departure in engine test data is caused by the presence of other sources, notably those generated internally but radiated as part of the exhaust field, jet temperature does have an effect on noise output. Experiments have shown that, at low jet velocities, increasing temperature tends to increase the noise generated, whilst at higher velocities the opposite effect is observed. There are several possible reasons for this phenomenon, all associated with the variations in the local speed of sound inside and outside the jet, and the existence of the hot–cold fluid boundary.[328-31]

However, since the temperature of the core jet on most engine designs does not vary significantly, the effects of temperature can usually be regarded as only a second-order factor in the design of an engine. The most important parameter is velocity, since this is the element in the engine design cycle that determines the overall significance of jet mixing noise. As a result, on zero- and low-bypass-ratio designs, where the core jet velocity can be anywhere from 500 to

600 m/s, jet noise is the overriding source. It is so dominant that other strong sources – for example, the compressor – are frequently inaudible in the high-power take-off phase of operation and are of equal intensity only at low engine powers used for approach to landing. The directivity of the jet noise pattern is such that it peaks towards the middle of the aft quadrant, with a gentle decay rate towards the forward quadrant and a more rapid decay rate to the rear of the engine into what is virtually a "cone of silence" within 10° or 15° of the jet exhaust axis. This region results from refraction effects in the boundary between the hot, high-speed jet and the cold atmosphere. Its presence can be observed quite readily when an aircraft overflies, banks and then turns directly away from the observer. The noise becomes almost inaudible as the observer is looking straight along the exhaust axis and into the back of the engines.

Figure 3.37. Jet mixing noise intensity versus jet velocity. (After Ref. 211, by permission)

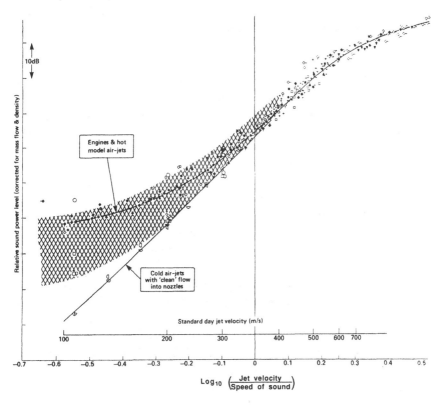

3.7.3   Flight effects

Because the intensity of jet mixing noise is a function of the velocity at which the jet mixes with the atmosphere, the noise level might be expected to fall as an aircraft accelerates from rest to become airborne, and the reduction felt all around the aircraft. However, for years, the experimental evidence suggested that there was a large reduction only to the rear of the engine and that ahead of it there was an amplification of the static level. Many experimental arrangements were used to simulate the flight condition in small engines and model jets. These ranged from an experimental hovertrain equipped with a small engine,[340-1] to a rotor arm with a tip jet[342] to jets in wind tunnels or secondary air streams.[343-9]

Eventually, acoustic wind-tunnel experiments using aerodynamically clean jets did reveal the full flight effect[320] and with it came the realisation that some amplification is due to convective Doppler effects, but that this had been overemphasised by the presence of noise from other sources that were not a function of jet velocity – for example, noise associated with shock, combustion, and other engine-internal sources. All these sources were, in fact, masking the real effects associated with the moving jet.[351-3]

3.7.4   The two-stream exhaust

Because jet exhaust velocity is intimately tied to the design cycle of the engine, which has to suit the mission of the aeroplane, there is little room for manoeuvre once the engine design has been frozen. With so little flexibility available to the designer of the early engines, much effort was directed at reducing jet noise by modifying the geometry of the final nozzle. The overall goal was to produce more rapid mixing and reduce the mean velocity in that region of the jet responsible for the dominant and most annoying sounds. The subject is further discussed in Chapter 4.

It was the low to mid frequencies that caused much of the problem, since they were generated at high intensities and are never much affected by natural atmospheric absorption. Therefore, they propagate over large distances and cause widespread complaints. Fortunately, despite having larger dimensions, the high-bypass-turbofan engine has greatly alleviated the jet mixing noise problem, because mean jet velocities have been reduced by as much as 50%. This change, with a V^8 velocity dependence, means a noise reduction of 24 dB at the same mass flow. Of course, "at the same massflow" means a reduction in thrust, but restoration of thrust by increasing the massflow only reduces the net change to 21 dB.

This enormous reduction is the fundamental reason why the turbofan engine has had such a marked beneficial effect on the airport environment. Moreover, with a much lower absolute jet velocity, fixed aircraft "forward speed" produces a bigger reduction in relative jet velocity. Even so, and despite the 20-dB improvement over earlier engines, the problem of jet noise has not been eliminated, and research continues accordingly.[360-3] In the 1970s, this research was extended to cover the case where the hot jet flow is ducted to surround the cold fan flow,[364-6] in the hope of lowering the noise generated. The work on this "inverted velocity profile" was motivated by interest in supersonic applications of modest bypass-ratio engines. The work fell by the wayside when fuel price increases made the SST less attractive for commercial operations.

On the subsonic, high-bypass-ratio front, the mixing pattern and associated noise-producing regions of the exhaust flow are somewhat more complex than the simple pure jet, that is, unless the hot core and cold fan flows are forcibly mixed by a mechanical device to give a uniform pressure and temperature profile at the nozzle exit. The two flows normally issue from separate concentric nozzles, with the added complication that one of the flows is two or three times hotter than the other and is exhausted at a 50% higher velocity. Frequently, it is also exhausted at a different axial position. This combination gives rise to several interfaces between hot and cold flows of differing velocity and a number of possible noise sources. In fact, theorists have utilised three or more source regions in trying to model the noise-producing process. The obvious separate source regions are illustrated in Figure 3.38.

Depending on the bypass ratio and the relative nozzle positions, the separate sources will vary in importance. Figure 3.38(a) shows that, for either modest or high-bypass-ratio engines, there is a choice of fan-duct length and that the configuration will have an impact on the relative noise levels of the different mixing region sources, as follows:

1. The fan jet will always mix with the atmosphere irrespective of the bypass ratio or nozzle position, and the mixing velocity will be the difference between the aircraft speed and the velocity of the fan exhaust.
2. The two jets will intermix at the difference between the core and the fan jet velocities.
3. Particularly in the case of short fan ducts of comparatively low-bypass-ratio engines, the fan jet will mix quickly and almost fully with the atmosphere and, further downstream, the

hot core jet will mix with what is virtually the free-stream environment. This mixing region will have the highest relative velocities in the total process, approaching the difference between the core velocity and the aircraft flight speed. At higher bypass ratios, with a "thicker" fan exhaust flow, the core flow will tend to mix in the fashion described in (2) above.

Finally, well downstream of the nozzle,

4. The mean jet flow (i.e., a combination of the core, the fan flow and some intermixed free-stream air) will mix with the atmosphere in a large-scale, but low-velocity, process.

The importance of these mixing processes can be appreciated by inspecting the different designs of turbofan engines that appear on a

Figure 3.38. Jet mixing noise regions behind a turbofan exhaust system. (After Ref. 211, by permission)

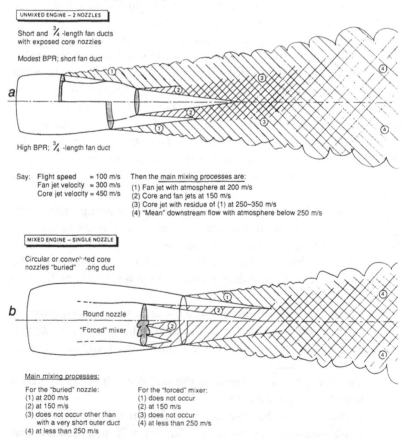

comparatively narrow range of aircraft. The original (modest-bypass-ratio) turbofan, the PW-JT3D, as fitted to the Douglas DC8 and the Boeing 707, appeared in two forms – one with a very short fan duct, with the fan flow being exhausted well upstream of the core jet, and another with a long fan duct. Today, the Boeing 747 has three competing high-bypass-ratio turbofan engines, all with different fan/core- engine exhaust-duct configurations. The early marques of the aeroplane with the PW-JT8D engine have very short fan ducts, whereas subsequent versions with the GE-CF6 and the RR-RB211 engines have much longer fan ducts, analogous to the $\frac{3}{4}$ duct of Figure 3.39(a). The latest -400 version of the 747 has an RB211 engine with an intermixer between the two flows, a long duct and a single exhaust nozzle. The smaller twin-jet Boeing 757 has three configurations, a short-duct

Figure 3.39. Typical turbofan exhaust configurations.

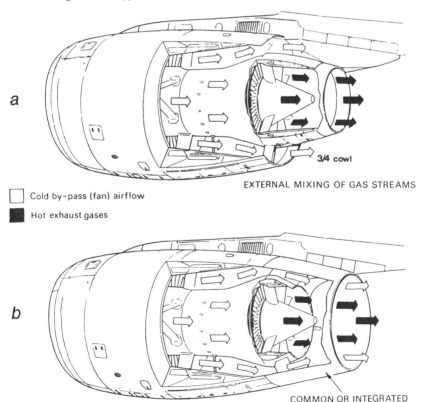

version of the PW-2037, and both short- and long-duct versions of the RB 211-535. In the long-duct version (Fig. 3.39(b)), the core nozzle is "buried" upstream of the final nozzle, and partial mixing is not aided by any mechanical device.

The mixing processes for the simple "buried" and the forced-mixed configuration are illustrated in Figure 3.38(b). Here, there is a higher degree of intermixing between the core and bypass flows before the presence of the free-stream air is felt. Hence, the mechanism labelled (3) in Figure 3.38(a), the highest velocity interaction, is now not present to the same degree as it is in the normal two-nozzle assembly. As a result, even the simple "buried" nozzle configuration tends to be somewhat quieter than the two-nozzle designs, and it also offers the possibility of further reducing mixing noise generated upstream of the final nozzle by incorporating acoustic lining in the exhaust duct, in the same way as a lined ejector works on a simple jet (see Chapter 4).

The presence of more than one important noise-generating area in the mixing region has been demonstrated experimentally from observations with source location instrumentation.[364] In the National Test Facility in England, the RB 211-535 buried-nozzle configuration mentioned above was reproduced at model scale, and an enveloping tertiary flow was introduced to simulate aircraft motion. The source location image (Fig. 3.40) indicated quite clearly the presence of at least two mix-

Figure 3.40. Use of a source location technique to identify two separate noise sources in turbofan jet mixing region.

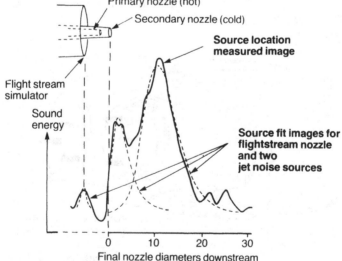

ing regions, one very close to the nozzle exit and another some 10 or so nozzle diameters downstream. By varying the ratio of secondary and primary jet velocities, it was further possible to establish that the upstream source was largely composed of high-frequency sound, whereas the downstream source constituted the lower portion of the overall frequency spectrum (Fig. 3.41). These observations could support the hypothesis that the high-frequency, near-nozzle source is caused by mixing of the fan flow with the environment, and that the downstream source results from the mixing of the "fully developed" jet. Variation of the secondary and primary velocity ratio only serves to alter the balance of high- and low-frequency components.

Theory is given further credence by source location results from full-scale engine tests. Figure 3.42 displays the 250-Hz data obtained from an engine with the most common $\frac{3}{4}$ cowl (two-stream) exhaust, like that shown in Figure 3.39(a). Clearly, at this frequency, there are two clear sources of jet noise whilst the internal noise from the fan and core engine are almost insignificant.

Figure 3.41. Spectral variations due to changing the ratio of primary to secondary jet velocity.

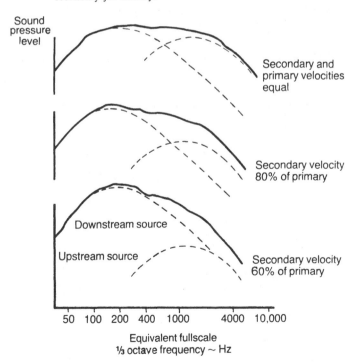

With all the different exhaust nozzle possibilities, it is not surprising that jet mixing noise should vary in intensity, spectral character and directionality across the range of exhaust geometries. Moreover, the internal jet-pipe geometry can have an effect on the noise generated in the mixing region, by virtue of the interaction of the aerodynamic perturbations they produce with the mixing process. That is why it has been very difficult to devise a generally applicable method of predicting coaxial-flow mixing noise.[956] Instead, most manufacturers utilise model exhaust systems to determine the precise changes in noise output from a "reference" configuration – usually a $\frac{3}{4}$ cowl or near-coplanar coaxial nozzle like that shown in Figure 3.39(a).

Note that all modern high-bypass engines have tended to produce a lower characteristic frequency spectrum because nozzle dimensions are larger and velocities are lower than in earlier engines. Frequency is, of course, inversely proportional to both jet velocity and general nozzle dimensions, or scale of turbulence.

When the first high-bypass-ratio turbofans entered service around 1970, there were complaints about low-frequency noise vibrating building structures, rattling windows and china and even causing nausea. Fortunately, such complaints were infrequent and the general reduction in jet noise across the board more than offset the limited effect of the somewhat lower characteristic frequency of sound.

Figure 3.42. Full-scale engine noise source location data at 250 Hz. (Source levels expressed relative to OASPL)

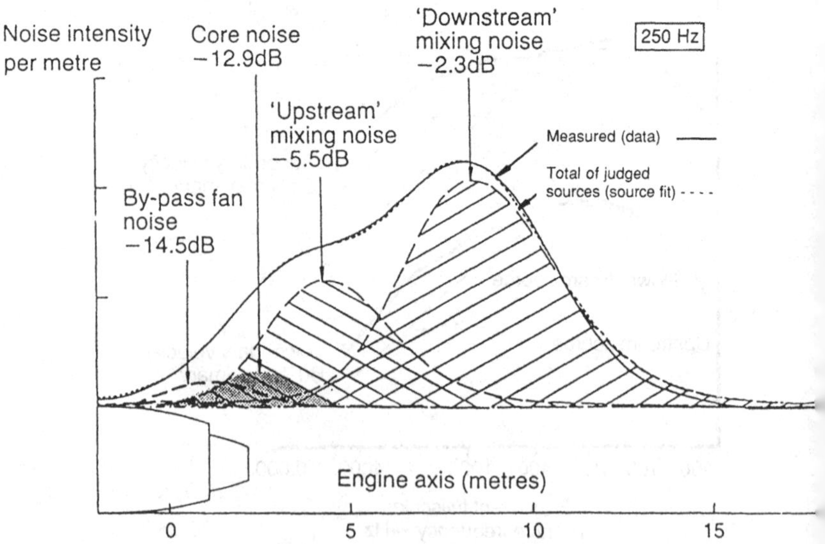

3.7.5   Other considerations

Often jet noise levels obtained from engine testing are slightly higher than those scaled from model testing and aircraft levels are sometimes higher than those noted in static engine tests at the same (relative) jet velocity. There are several possible explanations for these observations.

Firstly, it has been widely reported[370-3] that jet noise generated by aerodynamically "clean" models is amplified by the presence of high-amplitude discrete frequencies, not only at the frequency of the tone, but across an extremely wide range of adjacent frequencies. In an engine, tones are generated in the fan/compressor modules and in the turbine, any of which could account for an amplification of the mixing noise. Secondly, the interaction between the mixing region and turbulence in the exhaust flow from the engine[374] could, in theory, account for an increase in the mixing noise.

As regards the observations from aircraft flight measurements, the location of the engine with respect to the aircraft structure is critical,[375-7] and can give rise to new noise or can amplify jet and other engine noise enough to create the same effect as doubling the number of engines. The subject is further discussed in Section 3.11.

3.8   Noise of thrust reversers

Thrust reversers are devices that direct the flow from the jet exhaust in a forwards direction after the aircraft has touched down. The retarding force applied in this way augments the effect of the wheel brakes and, in slippery conditions, is often safer than the braking system alone.

The normal procedure on touch-down is to select reverse thrust at a negligible engine power and then to apply progressively increasing levels of power. For a short period in the braking process, the engines can sometimes produce almost as much noise as they do under full-power take-off conditions.[790] Fortunately, however, full engine power is not used in the reverse thrust mode, and the noise produced is heavily attenuated as it propagates over the ground.

Thrust reversers are complicated devices. They rely heavily upon the reliability of moving components and, depending upon the type of engine installation, take one of several forms. Figure 3.43 indicates the three most common types. Two rely upon internal air- and gas-deflection devices, whilst the third is external to the engine carcase. In Figure 3.43(a) doors move across the exhaust duct to deflect the flow partially forwards through a series of cascade vanes. In Figure 3.43(c) the same principle is applied to the fan flow of a turbofan engine. "Blocker" doors seal off the normal fan duct exit and cause the air to be deflected forwards through a series of cascades. In this case, the hot stream flow is not reversed, and is still allowed to provide forward thrust since it generally accounts for only 20% or less of total thrust.

The "target", or external "bucket" thrust reverser, is illustrated in Figure 3.43(b). Here, two externally mounted half-cylinder doors, which are normally flush with the power-plant nacelle, are translated rearwards and turned through almost 90° to provide a wide-angled V-shaped surface against which the jet exhaust impinges. A small

Figure 3.43. Methods of reversing engine thrust.

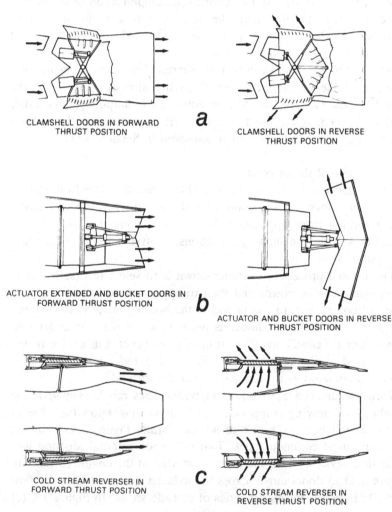

CLAMSHELL DOORS IN FORWARD
THRUST POSITION

a

CLAMSHELL DOORS IN REVERSE
THRUST POSITION

ACTUATOR EXTENDED AND BUCKET DOORS IN
FORWARD THRUST POSITION

b

ACTUATOR AND BUCKET DOORS IN REVERSE
THRUST POSITION

COLD STREAM REVERSER IN
FORWARD THRUST POSITION

c

COLD STREAM REVERSER IN
REVERSE THRUST POSITION

number of reversers actually split the nozzle in half to create the target doors – the Concorde employs this concept by using the variable secondary nozzle to provide the target doors (see Figure 5.3).

About half the world fleet uses the target reverser philosophy, and the remainder uses some variant of the cascade concept. Two examples, a wing-mounted engine with a cascade reverser and a rear-fuselage-mounted target type, are shown in Figures 3.44 and 3.45. Irrespective of type, they all create considerable noise when used. However, because of the complexity and high cost of the thrust-reverse mechanism and the unlikelihood that a practical method of suppression will be found, thrust-reverse noise tends to be accepted as a short-term necessary evil. The few controls that are in effect apply to the use of high-engine powers whilst in the reverse thrust mode, and at critical airports (usually at night) it is to be completely avoided in all but emergency situations (e.g., slippery/icy runway conditions or brake failure).

Fortunately, with the move towards high-bypass-ratio turbofans and the tendency to reverse only the fan stream, which is low-pressure-ratio, the noise produced has decreased over the years, roughly at the same rate as take-off noise. As high-bypass technology spreads throughout the operational fleet, the general level of noise from thrust-reverser usage should continue to come down.

Figure 3.44. The cascade thrust reverser.

3.9   Combustion noise and other engine sources

The complex turbomachine that makes up the modern aircraft engine can produce noise in a multitude of areas other than those already described. Few are relevant in the overall picture, but one or two of these bear consideration,[740-82] such as the combustion system, the bleed valves that offload high-pressure air from the compressor system into the bypass duct at compressor conditions that are "off-design", internal aerodynamic devices like the exhaust mixers and even the rough surface of the acoustically absorbent liners that are introduced to suppress internal noise. Not all of these are of great relevance, but they do merit some discussion.

A diagram of a typical combustion chamber is shown in Figure 3.46. Some engines have several combustion chambers arranged

Figure 3.45. The target thrust reverser.

around the annular compressor exit, whereas in others the combustion chamber itself forms a complete annulus, with a series of fuel injectors and igniters spaced around it. Irrespective of type, the combustion chamber can be divided into two basic zones. The primary zone is the area where the fuel injection and burning processes take place. Here, the gas temperatures are over 2000 °K, which is far too hot for entry into the turbine system. The secondary zone is the area in which hot gas is diluted with cold air to bring the gas temperatures down to a manageable level. Even so, turbine entry temperatures are usually in excess of the metal melting point but, thanks to cooling technology, they do hold together for thousands of hours!

To facilitate good fuel–air and gas–air mixing, the whole combustion process has to be extremely turbulent, and one would expect it to be a source of considerable broadband noise. Although the turbulence inside the combustion chamber does tend to be somewhat smoothed out owing to the contraction that occurs as it passes through the nozzle guide vanes at entry to the turbine, broadband noise generated in the combustion chamber does emerge through the final nozzle of the engine as a measurable constituent.

Some early research[740] dismissed the combustion process as irrelevant in the total noise observed to emanate from an engine exhaust. Moreover, despite all the work on jet noise in the 1950s and 1960s, it took a long while for researchers to determine that combustion noise was a recognisable component in the total sound signal that was being loosely attributed to the jet mixing process. It was not until the aerodynamic rigs that were being used for jet noise experiments were improved by smoothing the poor-quality flow and muffling the noise generated in the combustion system and pipework upstream of the

Figure 3.46. Typical section of a combustion chamber. (From Ref. 211, by permission)

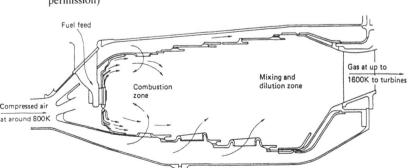

Fuel feed

Combustion zone

Mixing and dilution zone

Gas at up to 1600K to turbines

Compressed air at around 800K

nozzle, that it was recognised that jet noise did follow the theoretical V^8 variation down to low velocities. With this knowledge came the realisation that the significant departure from V^8 observed on engines running at low powers was due to the presence of internal sources variously labelled core,[741] tailpipe, combustion,[742] sometimes "nozzle-based" or even simply "excess" jet noise.[743] In hindsight, the confusion is understandable, for the spectrum associated with this source is very similar to that of the jet mixing noise, and it was (and still is) extremely easy to confuse the two. However, at angles corresponding to the refraction angle of sources propagating from inside the engine through the jet mixing region, it is possible to determine both a change in level and a separate spectrum shape associated with combustion and other internal sources (Fig. 3.47).

One reason that combustion noise has been ignored is that, under static conditions such as those pertaining on a model rig, jet mixing noise is considerably higher in level than it would be under flight conditions, where there is a much reduced relative (or effective) jet velocity. Thus, with the jet noise reduced under the flight condition, the combustion noise would be expected to stay at the same level and hence appear more important in the overall spectrum.

Fortunately, combustion noise has not yet proved to be a dominant problem on any engine design, despite being identifiable and measurable. Its origin has been the subject of considerable debate.[744-57] One view is that it results simply from the highly turbulent nature of the aerodynamic process in the combustion chamber although, as already pointed out, there is some smoothing in its passage through the nozzle

Figure 3.47. Identification of core noise in the $\frac{1}{3}$-octave spectrum of noise measured to the rear of a turbofan engine.

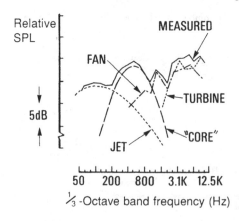

guide vanes. Another view is that combustion-related noise results from fluctuations in the temperature of the gas leaving the combustion chamber, which then interacts with downstream components, mainly the turbine, to produce noise. The final answer is by no means clear, although turbulent flames are known to produce high-energy noise (variously described as "roar" or "screech"), but the structure of an open flame is undoubtedly quite different from that in the metered and confined environment of a combustion chamber. Various attempts have been made to correlate combustion noise energy with the changes in pressure and temperature (and even flame speed) in the chamber,[975-79] and a variety of prediction procedures have been suggested. However, nobody has produced a definitive prediction procedure that takes into account all of the observed characteristics of gas turbine engines. Therefore, it is fortunate that combustion noise has not yet been important enough to necessitate fundamental changes to the design of the combustion system.[757]

3.10   The helicopter

Helicopter noise, a distant cousin of propeller noise, differs in that the rotor is aligned with the flight direction, rather than being normal to it. Consequently, few sources of lifting-rotor (and tail-rotor) noise are identical to those associated with the propeller. Furthermore, helicopter noise contains a significant additional source with a very characteristic sound. This is the phenomenon known as blade "slap" or "thump".[680-3] It occurs when one of the rotors passes through the wake or tip vortex created by another blade, especially on descent. It is also a feature of in-line two-rotor or "tandem" machines. Slap is not the only characteristic of helicopter noise, for each type produces its own distinctive sound, which is crisp in the case of two-bladed rotors but more like a thump or a bang in the case of large multibladed rotors. Add to the main rotor sound the noise from the tail rotor, which varies from mid to high frequencies, and no two helicopter types produce the same sound field.

Like propeller noise, helicopter rotor noise tends to grow in intensity and to change character as tipspeed and power increase, but there are significant asymmetric effects[689] that influence local tipspeeds and that are not a feature of the normal propeller.[684-7] Unless a helicopter is hovering, there will always be a substantial difference in relative (or helical) tipspeed between the blade on the "advancing" side and on the "retreating" side of the rotor disc plane, with respect to flight direction (see Fig. 3.48). For example, if the helicopter is cruising at a speed of 50 m/sec and the rotational tipspeed is 200 m/sec, the helical tipspeed variation, port to starboard, is half the basic rotational tipspeed. This will, of course, produce quite a marked difference in both aerodynamic lift and noise output, left to right, which would be even more extreme were it not for the cyclic pitch-control system that optimises the performance of the rotor as it undergoes sharp flow-field changes in one revolution. These changes can be very large. In the case of a helicopter with a high cruise speed, for example, the advancing blade may at times be close to sonic conditions, whereas the retreating blade may be close to stall. It is because of the high speed of the advancing blade that the noise output from the rotor assembly is asymmetric, and often preferentially directed ahead of the helicopter. The same effects occur, albeit to a lesser degree, with the tail rotor,[688] which has a top-to-bottom relative tipspeed variation.

Spectral analysis of the sound of a helicopter reveals a largely tonal signature, in fact, a combination of tones from the main and tail rotors. Main rotor tones are at a lower frequency than those of the tail, owing to the lower number of blades and slower rotational speed. These points are clearly evident in Figure 3.49, where the fundamental tone of the main rotor is less than 50 Hz and that of the tail rotor around 150 Hz.

Behind this apparently simple procession of main- and tail-rotor

Figure 3.48. Asymmetry in the aerodynamics of a helicopter blade.

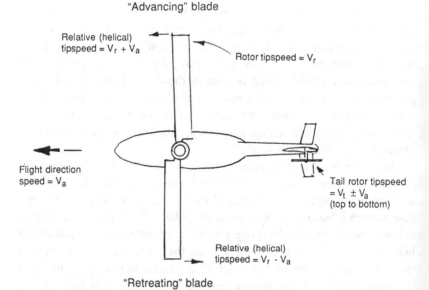

tones lies a multiplicity of sources. Submerged between the tonal signature is broadband noise,[690-1] from both the main and tail rotors, although it is the tonal sounds that normally dominate. Apart from the tones produced by the rotor blades, as a result of both steady and unsteady loading, there are the interactive effects between the rotors and fixed and moving structures. Two typical sources are illustrated in Figure 3.50 (3 and 4): the interactive effects between the main rotor wakes and the fuselage, and the tail rotor wakes and the vertical and horizontal tail structure. These are not to be confused with tip-vortex

Figure 3.49. Narrowband spectrum of helicopter noise, showing main rotor (R) and tail rotor (T) components. Harmonic numbers (1, 2, 3, ...) are given as subscripts. (From Ref. 211, by permission)

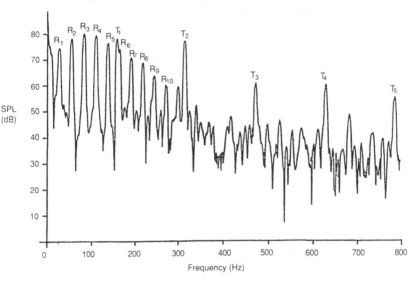

Figure 3.50. Typical helicopter noise sources.

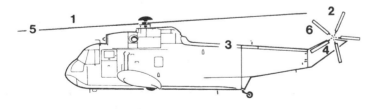

1&2   Main and tail rotor tones and broadband

3&4   Interactive tones (rotor/structure)

5&6   Tip vortex effects (slap); main and tail rotor

effects (5 and 6), which result in slap when the main rotor intercepts the vortex shed by the preceding blade, or in bubbling sounds when vortices shed from the main rotor are intercepted by the tail rotor. This latter effect exhibits a large harmonic content in the frequency spectrum.

All the foregoing effects are examples of how helicopter main- and tail-rotor noises differ substantially from those associated with the simple propeller. The propeller analogy is only applicable to the main rotor and tail rotor under hovering conditions, that is, with no vehicle speed to cause asymmetric tip velocities on the main rotor or nonaxial inflow conditions on the tail rotor. At low blade speeds, the mean steady loading can be used to predict the fundamental rotor tone and low-order harmonics using a propeller analogy but, when blade thickness effects become important at the higher helical tip Mach numbers, thickness noise increases rapidly and the propeller analogy breaks down. Similarly, unsteady loading due to interactions with preceding rotor blade wakes and atmospheric turbulence,[693] or due to the effects of cyclic pitch control, are much more severe than in the case of the simple propeller. Thus, quantitative prediction is extremely difficult.[692]

It is also important to note that broadband noise can be confused with genuine multiple-tone harmonics (or subharmonics), and that measurements may be further complicated by ground reflection effects. There is no easy method of predicting broadband noise produced from turbulence, either of a random nature or resulting from boundary-layer shedding at the trailing edge, and in the tip vortex. Most prediction techniques rely on empirical correlations. Blade slap characteristics, or interactions with main-rotor-shed vorticity, are also extremely difficult to predict, for large areas of an individual blade will undergo incidence changes for a short time when the rotor passes through a tip vortex. If the axis of the vortex is more or less parallel to the span of the blade that it intersects, the effect is at a maximum, for when the vortex axis is normal to the intersected blade, only a small portion of the blade is affected.

Thus far, it has not proved possible to eliminate blade slap, although the operational characteristics of the helicopter are important in determining whether the effect is minimised. As Figure 3.51 shows, the onset of modest slap and its development into intense slap are a function of vehicle forward speed and its rate of descent. These two factors determine whether the tip vortex generated by one rotor is intercepted by the following rotor; no two helicopters have the same slap characteristics with respect to forward speed and rate of descent.

It is possible to produce an individual plot of the type in Figure 3.51 for each type of helicopter, and use the boundaries to determine the optimum approach procedure. Unfortunately, noise certification employs a rigid approach, and calls for a fixed glide slope and a fixed forward speed. It makes no allowance for the obvious possibility that the vehicle might be flown in such a way that it could avoid severe, or perhaps even modest, blade slap.

In recent years, the formulation of noise certification standards for helicopters has been the subject of considerable international debate. On the one hand, it has proved difficult to measure and report noise levels in terms of EPNdB to the same levels of accuracy as with fixed-wing aircraft; on the other hand, opinions vary as to the subjective importance of blade slap, and how the recognised scales should be adjusted for such "impulsive" sounds.[172-3] A number of these outstanding questions have been explored in an international measurement exercise, the results of which are still being assessed.[695] The helicopter community awaits the outcome.

3.11   Installation effects

When a power source is installed on an aircraft structure, many additional noise effects are introduced above the basic ones produced by the power-plant and the airframe. The generation of

Figure 3.51. Helicopter impulsive noise (blade slap) boundaries.

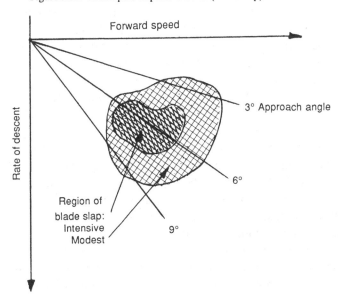

tones at rotor-blade-passing frequency as a result of installing a propeller power-plant close to a wing, or a helicopter rotor close to the
fuselage, are examples of installation effects. They all need to be
considered over and above the basic noise of the propeller or the
helicopter rotor.

There are many such examples, those mentioned above usually
being considered part of the generative process of discrete tone noise
associated with the propeller or the helicopter rotor. Others are less
tangible, less quantifiable and more likely to produce a few surprises.
Consider a typical commercial transport aircraft, and the installation of
the power-plants.

Figure 3.52 presents three typical installations of high-bypass-ratio
power-plants. Figure 3.52(a) shows a turbofan with separate fan and
core-engine exhaust systems mounted ahead of and beneath the wing
of a twin-jet aircraft; Figure 3.52(b) shows the installation of a high-
bypass-ratio, mixed-flow power-plant (single-exit nozzle) mounted on
the rear fuselage of a twin-engined executive jet; and Figure 3.52(c)
introduces a third common installation, an engine in the tail of a
three-engined aircraft. These installations demonstrate how various
reflective/refractive/shielding effects can arise as a result of the position
of the intake and/or the exhaust system with respect to major airframe
structures.

In the underwing mounted position, the noise radiated upwards
from the fan and core-engine exhaust system can clearly be reflected
back down from the under surface of the wing. Since the engine is
mounted at the front of the wing, not only does the noise from internal
sources stand a chance of reflecting downwards but also some of the
jet noise generated externally but close to the nozzle. Therefore, it is
hardly surprising to find that additional sources appear in noise data
measured on this kind of installation, compared to an isolated power-
plant. Model-scale experiments[760-1] have shown this quite clearly, and
Figure 3.53 indicates the effect of just this installation. Low-frequency
noise is augmented considerably, partly because new sound is generated as a result of having the flow close to such a large flat surface. Nor
is it surprising to find that, when the flap system at the rear of the wing
is deployed close to the exhaust jet of the power-plant, noise increases
considerably particularly if part of the flow impinges upon it.

A wing-mounted installation is always likely to produce more noise
with respect to the isolated power-plant, but the fuselage-mounted
installation of the twin jet in Figure 3.52(b) is different. The intake
of this engine is mounted almost above the main wing structure, and

there can be considerable shielding benefits beneath the aircraft. Conversely, the tailplane of the aircraft can act as a reflecting/scattering agent with respect to exhaust noise, which can measurably augment jet- and core-engine noise beneath the aircraft.

The wing-shielding effect was first noted many years ago in a comparison of the noise–time histories of a wide range of aircraft.[406] Figure 3.54 reproduces those data, showing the shortening of the duration of the noise signal due to a reduction in the forwards-radiated signal

Figure 3.52. Typical power-plant installations: (a) wing-mounted pod; (b) fuselage-mounted pod; (c) tail- and wing-mounted combination.

(a)

(b)

(c)

when the engines are mounted at the rear of the fuselage rather than under the wing.

Another installation effect associated with the fuselage-mounted power-plant location is seen in the noise radiated laterally. On the aircraft in Figure 3.52(b), there is a vertical "winglet" at the tip of the main wing structure. Even without such a winglet, the flow associated with the conventional wing, perhaps the wing-tip vortices, have been shown to have a scattering effect on noise radiated from the engine to an observer to the side of the aircraft; whether a tip winglet produces a larger or smaller effect is not yet known.

In the final configuration, shown in Figure 3.52(c), an engine is mounted in the tail of the aircraft. In this case, the noise from the engine inlet beneath the aircraft is almost completely shielded, whereas the noise from the exhaust behind the tailplane is free to radiate. This installation has, in effect, three "visible" engine exhausts, but only two air intakes.

Engines have also been installed in other ways over the years. Some examples of a completely different type of installation are the Vickers VC10 transport aircraft (Fig. 3.55) and the Lockheed JetStar business jet. On these aircraft, twin power-plants are mounted close together on either side of the rear fuselage to give a four-engined rear-mounted configuration. In this case, there is some merging of the adjacent exhaust streams, which modifies the noise radiation to the far-field. This is another type of installation in which wing-tip vortices have been

Figure 3.53. Installation effects in an engine mounted beneath the wing. (Crown Copyright – RAE)

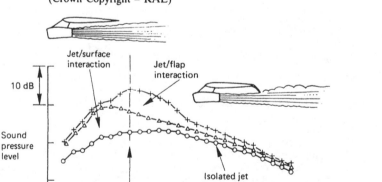

observed to produce measurable reductions in sound level to the side of the aircraft.

One aircraft, the VFW 614, has an extremely novel installation – above the wing. This is shown in Figure 3.56 and, as might be expected, there is a change in the field shape of the noise radiated beneath the aircraft when compared to that measured on a static

Figure 3.54. Observed differences in noise–time history for different power-plant installations.

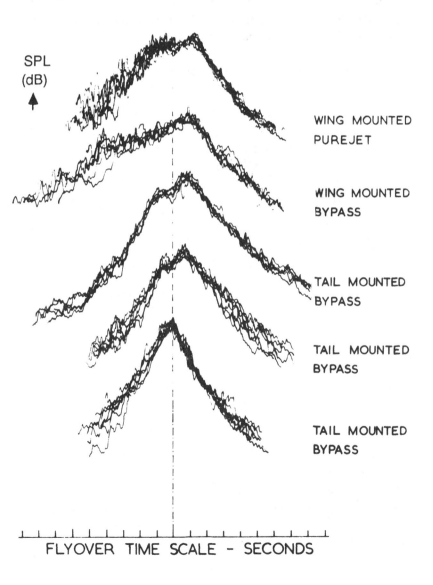

SPL
(dB)

WING MOUNTED
PUREJET

WING MOUNTED
BYPASS

TAIL MOUNTED
BYPASS

TAIL MOUNTED
BYPASS

TAIL MOUNTED
BYPASS

FLYOVER TIME SCALE - SECONDS

test-bed from the isolated power-plant. There is some shielding of the intake noise and some reduction of the noise radiated from the rear of the engine, both as a result of mechanical shielding by the wing and, when the flaps are deployed, aerodynamic shielding as a result of the diffused propagation of the noise signal through the turbulent wing wake.

In the late 1970s, there was a substantial theoretical and experimental effort to study wing and fuselage shielding, mainly with the objective of encouraging novel low-noise installations.[944] As a result, shielding was demonstrated to be a powerful way of reducing noise beneath the aircraft but, unfortunately, the aerodynamic and

Figure 3.55. The "paired" Conway engine installation in the Vickers VC10.

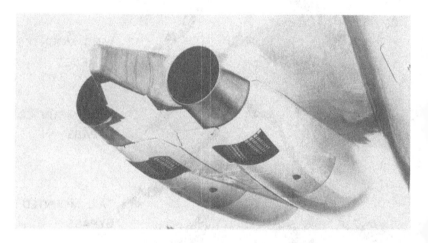

Figure 3.56. The overwing installation of the M45H in the VFW 614.

structure-weight penalties that would have to be absorbed in a long-range aircraft design have not made over-the-wing installations a commercially attractive proposition. Only one other exists outside the Soviet Union, a U.S. experimental aircraft, with many design objectives other than noise on the list, but it does have an over-the-wing exhaust system.

Some other experiments concerned with the general shielding concept followed the pattern of Figure 3.57, which shows a de Havilland Trident fitted with exhaust "scuttles". The purpose of this experiment was to evaluate the possibility of reflecting internally generated core-engine sources back up to the heavens, to the benefit of the population on the ground below. The experiment failed in that there was insufficient length in the scuttle devices to permit reflection of more than a small proportion of the total energy generated close to the nozzle exit. Any length added to the scuttles would have introduced an unacceptable drag penalty.

Not all such experiments are a total failure, however, and there is evidence in a number of aircraft to indicate the success of attention to installation features. The centre engine installations in the tail of

Figure 3.57. A de Havilland Trident fitted with experimental noise-reflecting scuttles.

modern trijets are extremely "clean", aerodynamically. Care was taken to reject proposals in the design schemes of the late 1960s to run the main tailplane structure up through the intake, ahead of the fan intake. This would have been disastrous in terms of additional fan (tone) sources. The trailing edge flaps on many aircraft are carefully designed to avoid both aerodynamic buffeting and flow effects that would introduce new noise sources; "blow-in" doors on the early turbofan intakes, devices that provide a measure of intake area variability, are no longer even considered. The variable nozzle of the Concorde is position-optimised to take benefit from directionality effects caused by squeezing two adjacent jets into a partial fishtail configuration. The list is surprisingly long.

3.12   Cabin noise

Cabin noise is produced in four principal ways. It is primarily the result of the combination of turbulence in the boundary layer on the outer surface of the cabin and engine noise transmitted either through the atmosphere or as a vibration through the airframe structure. It can, however, result from internal aircraft systems, particularly the cabin air-conditioning/pressurisation system and associated ducting. The worst cabin environments are those in the helicopter and in propeller-powered aircraft, where noise from the rotors is extremely difficult to control, being of low frequency and strongly transmitted via

Figure 3.58. Typical ranges of cabin noise.

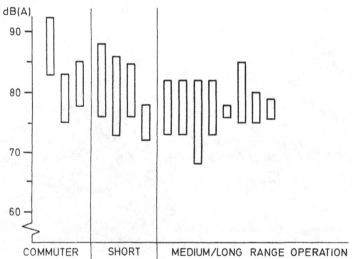

the airframe structure and through the atmosphere, particularly in the rotational plane of the propeller. This is one reason why propeller power-plants rapidly lost favour in the long-range market with the introduction of the jet engine. Not surprisingly, people's tolerance of noise is a function of the time that they have to withstand it, and in some aircraft, cabin conditions are tolerable only for a short period of time. Although there is a continuing debate over the most appropriate metric, cabin noise is usually measured in terms of the dBA unit, a good cabin environment being about 70 dBA, a bad one close to 90 dBA. Most modern passenger cabins are generally less than 80 dBA, and speech communication is good. The environment is usually best at the front of the aircraft, and it worsens progressively going rearwards as the fuselage boundary layer grows and the impact of engine (jet) noise increases. It is not merely a matter of coincidence that the first-class cabin is always at the front of the aircraft! Figure 3.58 illustrates that a medium- to long-range aircraft cabin environment is substantially quieter than that of the commuter and short-range aircraft, and that the noise level varies within the cabin by some 10 dBA or so.

Methods of controlling cabin noise, either by containing the noise at its source or by modifying the fuselage structure or cabin furnishings, all tend to reduce the performance of the aircraft and increase the cost of operation. Performance is lost either as a result of design action that suppresses noise in the engine, or because extra weight has to be carried in the fuselage to damp the vibration signals and sound-waves that are causing the problem.

Apart from occasional unusual transmission characteristics, the modern turbofan power-plant contributes little to cabin noise. This is because the airborne noise is well down, jet velocity and resultant mixing noise being low, and because vibration levels from turbine engines are normally of a low order. This may not be the case with the open-rotor concept if it reaches fruition in production aircraft, since is it not the case with the helicopter rotor and the propeller today. Close attention has to be paid to structural vibration damping of signals transmitted from the power-plant and resulting from rotor-wake interaction with parts of the fuselage (e.g., the wing in the case of a propeller mounted ahead of it). In the plane of the propeller, where blade-passing frequency and harmonic signals are of very high intensity, noise-control concepts vary from active mechanical damping of selected discrete frequencies to the addition of mass to cover wider frequency ranges. Consideration has also been given to using a technique of airborne "active" sound cancellation within the cabin environment,[890] through systems that detect the noise signal as it is generated and then cancel it by playing an out-of-phase signal through a series of loudspeakers. Thus far, however, this form of cabin noise control is still in the experimental stage. (It is a standing joke that, if active cancellation is fully successful, it must eliminate all conversation and that, if the system experiences a power failure, the passengers are likely to die of shock when all the noise is instantaneously restored!)

Fortunately, there are several well-established methods of cabin noise control and, since their use is intimately tied in with the design of the aircraft structure and the installation of the power-plants, it is a subject that requires separate detailed study. Reference material may be consulted as a start to any further pursuit of the subject.[891-4]

3.13   Military aircraft noise

Some of the noisiest aircraft ever produced are owned by the military. In particular, fighter aircraft are required to fly at high speeds and, like the Concorde, require an engine cycle that will provide a rapid thrust response over a wide range of operating speeds. Since large-diameter installations would result in far too much aircraft drag, the power-plant should have a low frontal area. This, and the requirement for high-speed operation, usually means that the engine cycle is either a straight turbojet or a low-bypass ratio. Since many of the civil engines now in service are developments of military power-plants, it follows that the noise sources are identical. It does not follow, however, that the same degree of attention is paid to noise control. Most military aircraft are on "standby" should the need for active service materialise. Consequently, any thoughts of noise control that involve performance loss are stillborn and, since even the use of acoustic lining to suppress compressor noise would be a weight penalty, no noise-control features appear in any purpose-designed military aircraft.

Noise, or, more precisely, a high-intensity sound field, is important in the structural integrity of military aircraft. With high-pressure-ratio engines having exhaust velocities of well over 600 m/sec, and many of them reheated to even higher velocities, the airframe structure in the proximity of the jet exhaust can suffer high excitation levels. Although some civil aircraft, particularly those powered by engines with high-velocity exhausts, have experienced acoustic fatigue failures,[304] the problem is far more severe in the case of military aircraft. This is largely due to the compactness of the engine installation and its integration into the airframe structure. In the typical military fighter, the power-plant, or system, takes up virtually the whole of the fuselage section whereas, of course, a civil aircraft relies upon its fuselage to carry people and freight, and its engines are mounted further from the fuselage; for example, beneath the wings. The only military aircraft that are analogous to the civil sector are those used for carrying troops, surveillance or refuelling. Many of there are actually civil aircraft that have been adapted to meet the special requirements of the military, but retain the civil noise-control features and have no additional acoustic fatigue problems.

Because the engine cycles of military combat aircraft are well outside the norm for the modern civil engine, there has been additional research into the noise from high-velocity jets, which continues as engine pressure ratios increase to accommodate the extra demands placed on each new design of aircraft. However, provided that the source spectrum can be defined with a reasonable degree of accuracy, methods exist for predicting the impact of acoustic fatigue[946] and, although these may have to be extended as military engine-pressure ratios and exhaust velocities rise, they have proved to be an adequate design tool.

3.14 Other aircraft noise sources

There are other aircraft noise sources not specifically dealt with in this chapter, which may or may not be of consequence in the overall airport noise problem. These include auxiliary power units, air and hydraulic power systems and "funny-sounding noises".

3.14.1 Auxiliary power units (APUs)

These units supply the essential needs of the aircraft whilst it is on the ground at the airport and the main engines are not operating, or when no external power supply is available. They are small gas turbine engines, normally mounted in the rear of the fuselage, which provide an audible source of noise to those in the vicinity of an aircraft that is being prepared for a flight. The noise of these units is normally only of concern to ground crew, and the manufacturing and operating industry have defined acceptance levels and methods of measurement.[909] APUs are not considered particularly important in the overall airport community noise problem, although some airports do try to shield parked aircraft from nearby communities.

3.14.2 Air and hydraulic power systems

These systems are of no consequence in the overall airport noise scene but they can disturb passengers. Air and electrical power motors, hydraulic pumps and other system-actuating devices can cause considerable noise inside the cabin, which, to a first-time flyer, may not only be unexpected but may be of surprisingly high intensity. Considerate airlines will either warn the passenger of characteristic noises in the preflight safety demonstration or in the seat-pocket literature.

3.14.3 FSNs

Funny-sounding noises, or FSNs, are an everyday feature of life in the aircraft business. FSNs may be reported by pilots, service engineers, test engineers and even passengers aboard scheduled flights.

FSNs represent a change of state – that is, the abnormal. Most frequently, they reflect a change in the steady background noise that is triggered by some unpredicted mechanical or aerodynamic event. There are many good examples, and in a number of cases the diagnosis of an unusual sound has resulted in the detection of and solution to an incipient development problem. Some examples follow:

> At least one high-bypass turbofan engine has suffered from an FSN during low-power operation on the terminal ramp. A rumbling was noted at certain power conditions, which approached deafening proportions to passengers who had to board whilst an engine was running. The cause was an instability in the combustion process which, if it had been allowed to continue, would have eventually caused a mechanical failure in the combustion system.

> One aircraft used to play a signature tune each time it took off. Its signature tune took the form of a piercing "hoot", which was eventually traced to a cavity resonance in the combustion system. This hoot was by far the most powerful sound on what was a noisy straight turbojet-engine cycle.

> An FSN was reported by passengers and crew on an executive jet on which the exhaust nozzle configuration had been changed. This change of state was simply caused by a change in spectral content and directivity of the jet mixing noise behind the engine. The fact that attention was drawn to the FSN prompted a technical investigation, which showed that the change in mixing noise would have caused an unexpected structural fatigue problem in the tailplane had it not been noticed early on.

Many FSNs are associated with combustion systems, perhaps the most notorious being a "buzz" from unstable burning in the combustion chamber or in a military reheat/afterburning system. These are analogous to the first example in this list, and have to be cured either by changes to the geometry of the combustion chamber or the mode of fuel supply to ensure efficient, smooth burning, or by acoustic damping in the exhaust ducting.

One modern turbofan-powered aircraft plagued its makers with a low-frequency "howl" when the flaps were deployed on landing. Although of no significance in the community noise context, it did cause some concern to passengers. Modern flap systems are highly efficient in increasing the lift forces on the wing, but they are riddled with slots and gaps that are ideal aerodynamic noise generators. The cure to the noise might well degrade their performance.

Wheel bays are a frequent source of FSNs – usually low-frequency discrete tones. The solution usually lies in improved sealing around the undercarriage door cavities.

FSNs can be used as a diagnostic tool. Fan blades exhibiting flutter will produce a noise signal at the natural frequency of vibration, which can be detected and used as an advance warning of impending failure. Incipient stall in a compressor can be detected by changes in the broadband level. Noise sensors are frequently used as condition monitors whilst internal stresses and strains in structures can be examined by measuring acoustic emissions.

The above are but a few examples of FSNs. Most are of little concern and merely reflect a small change in the normal steady-state environment. Some, however, warn of a genuine and developing problem – some signify that the problem has fully developed. FSNs cannot be ignored.

4

Power-plant noise control

Over the past thirty to forty years, a vast knowledge of aircraft noise-control techniques has been acquired. Some of the findings have contributed to the improved airport-noise climate of today; the less successful and the failures have taught both useful and bitter lessons. This chapter concentrates on success – success in the field of the jet engine, for it was this propulsion system that really started the serious noise problem.

For a broader understanding of the problem, the specific means of controlling aircraft power-plant noise should be discussed in relation to design philosophy, which, in turn, is related to either date of concept or aircraft mission requirements. The following sections deal with the range of power-plants in service in two convenient categories, low- and high-bypass-ratio types.

4.1 Suppression of the early jets

As explained in Chapter 3, it is the basic cycle and the maximum thrust level of the engine that determines the level of jet noise produced. The first generation of pure jets and, to a large extent, the low-bypass-ratio engines that succeeded them, all had extremely high exhaust velocities, which caused high levels of jet noise.[301] At full power, supersonic exhaust flows with velocities of up to 600 m/sec were not uncommon and the characteristic crackling[310] and tearing sounds of the mixing noise were augmented by the presence of shock-associated noise.[306]

Without today's turbine cooling technology, there was no opportunity whatsoever to reduce the jet noise in these engines by modifying the engine cycle. The modest properties of the "hot end" materials limited operating temperatures in the core engine and hence the amount of energy that could be extracted by the turbine. As a result, the jet velocity could not be reduced, since it was controlled by the thrust requirement via the available total airflow, and jet noise had to be reduced by other approaches. These took the form of mechanical devices aimed at modifying the aerodynamic structure of the mixing process and controlling the energy dissipated as noise.

The late 1950s and early 1960s saw an explosion of experimental work directed at understanding both the aerodynamic and acoustic processes in the jet mixing region. This was supported by a wealth of theoretical work. In fact, it became fashionable to develop theoretical explanations for observed experimental phenomena but, unfortunately, all the theoretical "solutions" of the era contributed very little to noise suppression of the early jet aircraft. The various suppressor designs that emerged in Europe[820-1] and the United States resulted from the efforts of experimentalists who believed that subdivision of the normal circular jet into numerous elements would promote rapid mixing. This, in turn, would cause a reduction in the velocities of the large-scale turbulent eddies that were responsible for the high-energy, low-frequency noise that was most objectionable (Fig. 4.1).

Figure 4.1. The jet-suppressor principle.

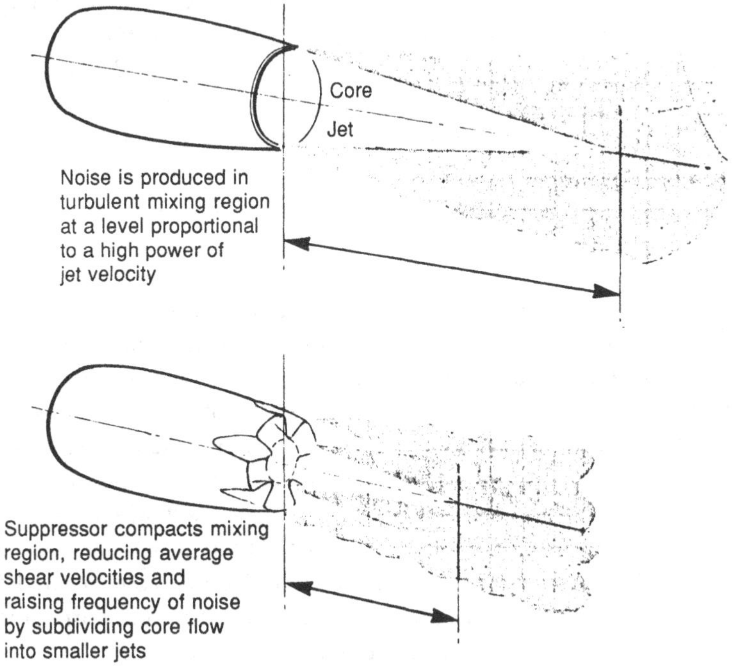

Core
Jet

Noise is produced in
turbulent mixing region
at a level proportional
to a high power of
jet velocity

Suppressor compacts mixing
region, reducing average
shear velocities and
raising frequency of noise
by subdividing core flow
into smaller jets

Although all noise suppressors do reduce the level of low-frequency noise to a certain degree, their main benefits are that they minimise shock noise by disturbing the shock structure in the supercritical jet, increase the mixing rate (and in doing so enlarge the "cone of silence" behind the jet) and, what is most important, they raise the characteristic frequency spectrum of mixing noise. The frequency shift comes from the reduction of the important dimensions associated with the mixing process, which results from the subdivision of the exhaust flow into several smaller jets. The increase in mixing-noise frequency then allows nature to take its course, for the atmosphere absorbs sound to an increasing degree as frequency is increased.[229] Consequently, a jet aircraft with several hundred metres between its "suppressed" exhaust and the population benefits from several effects (see Fig. 4.2). Firstly, there is the low-frequency noise reduction resulting from the nozzle change. Secondly, there is the natural high-frequency attenuation of the atmosphere and, thirdly, the loss of high-frequency shock noise negates the frequency increase caused by the suppressor. Even though the attenuation characteristics of the atmosphere do vary considerably with changes in both temperature and relative humidity, the amount of high-frequency absorption can be several tens of decibels under average climatic conditions.

At very high jet velocities, the loss of shock noise is as important as the reduction in low-frequency energy, since it falls in that region of the spectrum important in the calculation of the perceived noise level. At subcritical jet conditions, with no shock noise to control, the usefulness of a suppressor nozzle soon disappears; in fact, under approach conditions, a suppressor can actually increase the computed perceived noise level, since the high frequencies are not absorbed in the short distance between the aircraft and the people beneath its low-angle glide slope.

The approaches taken by different manufacturers to the practical problem of subdividing the jet, with minimum performance loss, were all variations on the rapid-mixing theme. Some favoured the development of the exhaust into a number of small tubular outlets, whilst others "indented" the nozzle exit-flow profile by inserting chutes or corrugations around the periphery. The earliest published work on jet suppression dates back to the early 1950s, a period when pure jets were being developed for commercial application from their military forebears. The literature reveals that work in Europe concentrated on nozzle designs that, broadly speaking, were based on the corrugation of the outer profile of the exhaust nozzle (Fig. 4.3); work in the United

States echoed this philosophy and also extended work into the multi-tube arena. A range of exhaust suppressors actually reached the stage of passenger-carrying service, both of tubular and lobular form and, in one case, the principle was extended to the use of a secondary ejector.

Figure 4.2. Spectral changes resulting from fitting a suppressor nozzle to a supercritical jet.

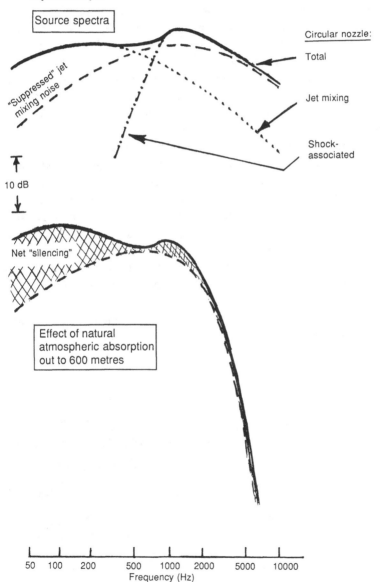

All the early work concentrated on noise suppression at fairly high jet velocities. Data were viewed in terms of the impact of what were believed to be the major relevant parameters. Generalised correlations were produced expressing the amount of noise reduction as a function of geometric changes, for example, the number and depth of corrugations, usually at a fixed jet velocity of around 500–600 m/sec (Figs. 4.4 and 4.5). Rarely were the different influences of shock and mixing noise addressed separately, with the result that correlations were extremely general, and led to some outrageous concepts. Although all the concepts had the same objective (i.e., to produce maximum low-frequency sound reduction), the aerodynamic performance was extremely variable when the different design features were introduced. Performance loss arises from two effects: the external "scrubbing" drag resulting from air flowing over the more complex geometric shapes that the suppressor introduced at the back of the engine nacelle and the internal degradation of thrust by losses associated with the more complex gas passages. It does not take an experienced aerodynamicist to appreciate that the internal thrust loss and external drag effects exhibited by the types of suppressor shown in the photographs of Figures 4.6 and 4.7 differ markedly; or that the sheer weight of the in-line eight-tube device of Figure 4.8 is a substantial penalty against the performance of the aircraft! Not shown is the ultimate in tubular suppressors that one organisation tested – which comprised 259 identical tubes. The weight of the device probably approached the weight of the engine!

In practice, performance losses were contained to an acceptable level, because "acceptable" was defined as what was necessary to

Figure 4.3. Two experimental corrugated nozzles of the 1950s.

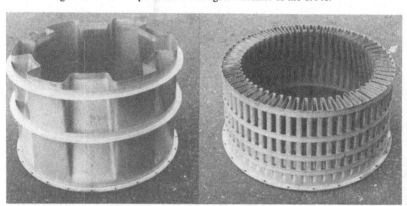

Figure 4.4. Early indications of spectral changes resulting from varying the number of suppressor nozzle corrugations.

DEPTH OF CORRUGATIONS 2·65 INS

N° OF CORRUGATIONS

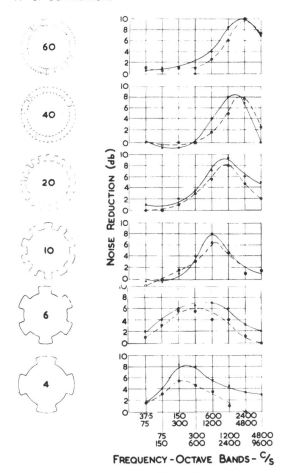

NOISE MEASURED AT 50 FT RADIUS
●——● 15° FROM JET AXIS ○— — —○ 30° FROM JET AXIS
JET VELOCITY 1800 FT/SEC

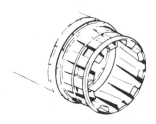

Figure 4.5. Early correlation of noise reduction versus "area ratio" $\bar{A}$ = (total gas + air area/gas nozzle area).

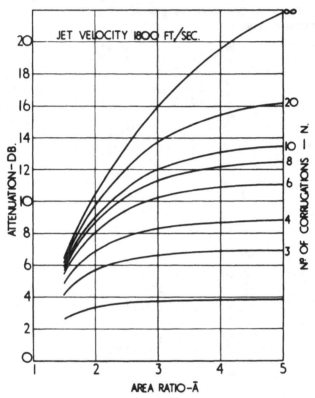

Figure 4.6. The original six-chute nozzle for the 1950s Comet aircraft.

make the aircraft tolerable at the airports it had to use. For most suppressor designs, the exchange rate of noise reduction against take-off thrust loss, engine fuel consumption increase at cruise conditions and the effects of increased weight and external drag are as shown in Figure 4.9. At around $\frac{1}{2}$% penalty in each case, for a peak-angle noise reduction of about 5–8 PNdB (or 3–5 EPNdB in flight), the effective penalty worked out to about a 2–3% increase in fuel burned for each mission. The equivalent figure on a modern turbofan installation is nearer a third of the penalty for three times as great a noise reduction, or an order-of-magnitude improvement. However, in both cases, at greater levels of noise suppression the penalties soon begin to escalate, no matter what the basic design looks like. In the case of the exhaust suppressor, this results from the increased complexity of the nozzle geometry, and the attendant rapid rise in aerodynamic losses. Later work, in the 1970s,[8,37-9] showed that it was necessary to go to the complexity of a translating lined ejector-suppressor system, operative only on take-off and approach, to keep drag losses at climb, cruise and descent low enough to get more than a 10-dB noise reduction for less than a 5% increase in fuel consumption – that is, to maintain the performance/noise trade-off rate of the simpler but less effective suppressors, established at about 1 dB for $\frac{1}{2}$% fuel burn penalty, at much higher levels of suppression. (See also Figure 5.9.)

Some of the suppressors that did reach commercial operation are

Figure 4.7. Experimental chuted eight-tube nozzle – "convoluted tubular technology".

shown in Figures 4.10, 4.11 and 4.12, whilst the only commercial application of the ejector concept to the DC8 aircraft is shown in Figure 4.13. In this case, an equally important feature was that the translating ejector "sleeve" contained doors that allowed the assembly to act as a thrust reverser. In fact, this device was far more successful in controlling engine thrust direction than it was in reducing noise.

Aside from the novel use of the system to reverse thrust, it should be appreciated that when an extra remote cowl, or ejector, is placed around the final nozzle of an engine, it can both augment thrust and reduce noise. Thrust is augmented when extra air is induced into the ejector and then is partly mixed with the exhaust gas and accelerated

Figure 4.8. Experimental in-line eight-tube nozzle.

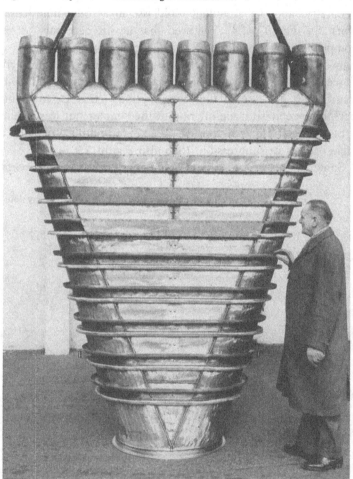

Figure 4.9. Early suppressor performance penalties.

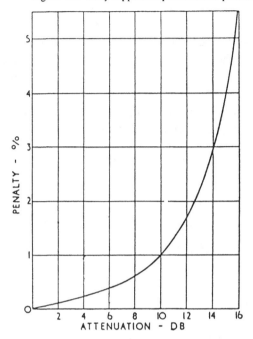

Figure 4.10. Eight-lobe nozzle on Conway engines in 1960s Boeing 707-420.

before being expelled through what is now a new and secondary nozzle, at the rear of the ejector. Since the added massflow causes velocities in the jet downstream of the ejector to be lower than they are at exit from the engine nozzle, there is some basic reduction in noise, but it is limited to the lower frequencies. The high-frequency noise generated by eddies in the mixing region close to the nozzle is largely unaffected, and it is free to radiate both forwards and rear-wards from the inlet and the exhaust of the ejector. In fact, because of the increased periphery of the rapid-mixer nozzle (that replaces the

Figure 4.11. Deep five-chute nozzle on Spey in 1980s Fokker F28.

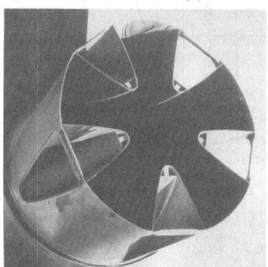

Figure 4.12. Twenty-one-tube nozzle for 1960s Boeing 707.

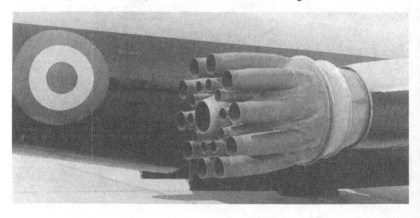

normal circular nozzle), the high-frequency noise is increased and, as has already been explained, it is only the effect of atmospheric absorption over large distances that produces the perception of reduced noise. However, if the ejector is lined with acoustically absorbent material, as shown in Figure 4.14, there is a much greater reduction. The liner absorbs much of the high-frequency noise generated close to the nozzle exit, by as much as 10 dB, and the gross effect is a perceived noise level reduction that is more than twice that achieved with simple suppressor nozzles. Moreover, whereas the benefit of most suppressors diminishes with reducing exhaust velocity, a lined ejector can be designed to produce near-constant attenuation over the extremely wide operating range of the power-plants, because high frequencies are absorbed in the ejector barrel.

Figure 4.13. Ejector–suppressors on 1960s Douglas DC8.

Figure 4.14. Principle of lined ejector–suppressor. (From Ref. 211, by permission)

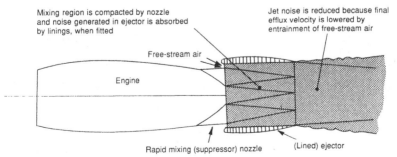

Mixing region is compacted by nozzle and noise generated in ejector is absorbed by linings, when fitted

Jet noise is reduced because final efflux velocity is lowered by entrainment of free-stream air

Free-stream air

Engine

Rapid mixing (suppressor) nozzle

(Lined) ejector

During the 1970s, when noise legislation was applied to the later versions of low-bypass-ratio-powered aircraft, there was renewed interest in the ejector principle. It also received consideration in studies of advanced supersonic transport concepts. An aircraft that is going to cruise at supersonic speeds needs an engine cycle that has high exhaust velocities at cruise conditions, and unless the engine cycle can be varied to produce a much higher massflow (and hence lower exhaust velocity) at low aircraft speeds, supersonic transport take-off noise is always going to be a design limitation.

One tangible outcome of all this work was that two flight development programmes began testing a fully lined ejector, firstly, in the United States[837] and then later in a joint U.K.–U.S. programme flown in England.[838-9] In the latter case, when fitted behind an RR Viper engine on an HS-125 executive jet (Fig. 4.15), a series of "fixed" ejectors and nozzles demonstrated the full potential of the system. Unfortunately, with the development of the high-bypass-ratio subsonic engine and the loss of interest in the SST when fuel prices rose dramatically, the technology was never exploited commercially. However, pressure is still being exerted on older, noisier aircraft, and the ejector concept is once again being considered for those aircraft that have a lot of structural life remaining but that are finding it difficult to live alongside low-noise turbofans. Thus far, however, the nearest a lined ejector has come to seeing service is in a full-production specification and initial manufacture for the BAC 1-11, but U.K. production of the aircraft was terminated earlier than expected.

Nevertheless, because noise certification standards were made applicable to older aircraft, the 1970s also saw a general upsurge in efforts to control jet noise. This coincided with the work on the high-

Figure 4.15. Experimental lined ejector–suppressor on flight/test vehicle.

bypass-ratio turbofan. Consequently, the hushkits that were developed for pure jets (in executive aircraft)[826] and low-bypass-ratio commercial fleet engines like the PW-JT8D and RR-Spey all benefited from the new work, particularly in the area of noise-absorbent liner technology. Although the Spey had a considerably higher exhaust velocity than the JT8D and required an exhaust jet suppressor, JT8D applications were made "legal" simply through the use of acoustic liners, which reduced internal turbomachinery sources. The subsequent development of the "refanned" JT8D,[845] which reduced jet noise by increasing the bypass ratio and hence lowering the jet velocity, represented a more substantial advance. The later RR Tay[846] is a similar refan of the Spey, but has a higher bypass ratio and lower levels of exhaust noise.

Even on a low-bypass power-plant, a tremendous amount of noise energy propagates both out of the inlet and down the exhaust duct to radiate with the jet noise, as illustrated by a test conducted on a Spey in the 1970s. Apart from a modified lobular exhaust suppressor, the rear end of the engine was fitted with a 200-cm long, 15-cm deep, acoustically absorbent tailpipe designed to remove almost all the noise energy transmitted down the assembly. As shown in Figure 4.16, the experiment was successful and, at the near-idle power condition illustrated, the jet mixing noise "floor" was achieved, as sound levels from the baseline engine were reduced by 10–20 dB at all frequencies, particularly at frequencies above 500 Hz.

Figure 4.16. Demonstration of the amount of internal noise emerging from the rear of a low-bypass-ratio engine at low engine power.

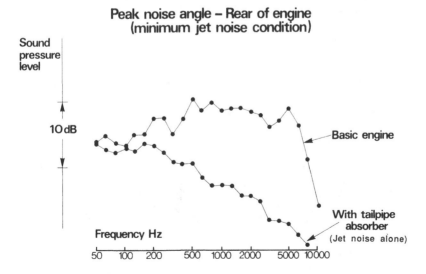

The energy in the core-engine flow arises from such sources as the combustion system, turbulence created in the forced mixing region between the bypass and core-engine flows, the turbine, and even noise propagating down the bypass duct from the HP compressor. Hence, liberal use of acoustic liners, even in a low-bypass engine, can have great impact on noise control. Most hushkits have utilised absorbent material in the intake and in the jetpipe, and some even considered it for the bypass duct. A typical hushkit arrangement is shown in Figure 4.17.

4.2   Noise control in the modern turbofan

The area that usually attracts the most attention on the modern turbofan is the low-pressure fan, but its associated core-engine compressor can also be important. The key sources that require attention in the fan system are the many tones, although the more randomly generated broadband noise can also be significant. Tones fall into several categories, but those that require attention at the design stage are the supersonically induced buzzsaw tones, intake flow–distortion interaction tones, rotor–OGV interaction tones and the multiplicity of interaction tones that occur in the leading stages of the core-engine compressor.

Buzz noise is controlled in three ways: by designing the leading edge of the blade to produce the minimum shock strength; by ensuring that the manufacturing process yields blades that are as identical as possible, and that remain so under rotational conditions; and by using acoustically absorbent material in the intake duct (covered in Section 4.3). Unfortunately, however, the ravages of service life usually mean

Figure 4.17. The 1970s hushkit used to control noise from the low-bypass-ratio engine. (From Ref. 211, by permission)

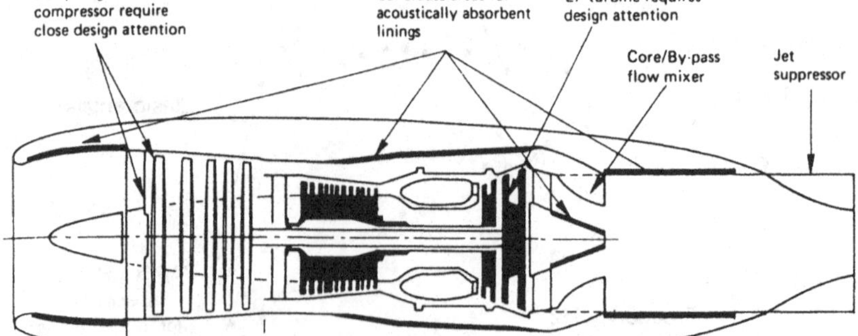

that no two blades are identical after they have been in service for some time, even if they are almost identical at the manufacturing stage. "Foreign object damage" (FOD), caused by the ingestion of ice, birds, runway debris and even hail and rain, will progressively affect performance of the blade so that it no longer meets the high standard conceived in the design.

Nevertheless, the design of the blade is important, for the more rigid the basic blade can be made, within the constraints of weight control, the less the chances that foreign objects will damage the tip-region profile and thus make buzzsaw noise worse. A high proportion of fans in service are of sufficiently high aspect ratio (length : width, or span : chord) and susceptible to vibration excitation that they require midspan vibration dampers, or "snubbers", which lock adjacent blades together to form a fully interconnecting ring (see Fig. 4.18). Indeed, some blades are so flexible and vibration sensitive that they require two rows of dampers and are even more liable to produce buzz.

In the subsonic régime, the important tones are those resulting from the interaction between fan-blade wakes and the downstream guide

Figure 4.18. Comparison of high- and low-aspect-ratio fan blades: (a) high-aspect-ratio blade with midspan snubbers; (b) low-aspect-ratio or wide-chord blade without snubbers.

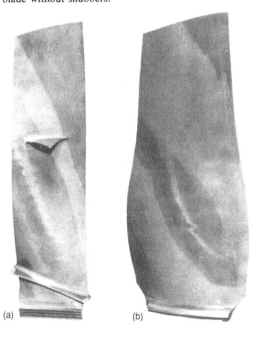

(a)   (b)

vanes in both the fan duct and the core engine. There may also be some interactions with the early stages of the core engine, which are responsible for "sum-and-difference" tones. As indicated in Figure 4.19, the interaction between the fan and tone outlet guide vanes is normally controlled by selecting the number of blades and vanes needed to "cut off" the fundamental tone, and/or ensure that significant modes are propagating counterclockwise and can take advantage of "rotor blockage" effects.[410-11] The outlet guide vanes will also be well spaced behind the fan rotor; typically, they have a gap of about twice the tip chord of the fan, so as to avoid potential field interaction and ensure that the blade wakes have decayed enough to generate a low level of tones. Similar attention has to be paid to the entry conditions into the core-engine compressor; otherwise this can be the dominant "front-end" source.

Another fundamental requirement of the design of the fan is that the entry airflow be aerodynamically "clean". For this reason, intakes are nearly always round and no longer feature devices such as "blow-in" doors, sometimes used to increase the flow capacity at high-power/low-speed conditions. Inlet guide vanes are also to be avoided; so are intensive arrays of aerodynamic sensors. All such obstructions produce disturbances that are sensed by the fan as cyclic changes in lift, which

Figure 4.19. High-bypass-ratio engine "front-end" noise aspects.

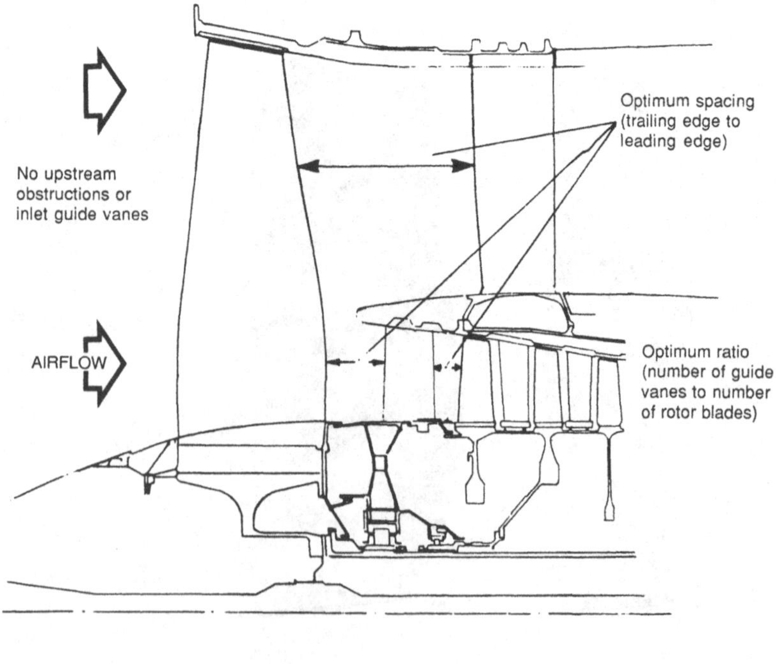

gives rise to discrete tones. Similarly, obstructions downstream of the fan can produce pressure-field disturbances that are sensed upstream by the fan and also produce tones. One such obstruction is the interservices pylon extension that crosses the fan duct and carries fuel to the engine and air and electrical power offtakes from it. Figure 4.20 shows how the aerodynamic design of one pylon was modified to reduce fan noise (and vibration) when the original design had been shown to induce a significant pressure-field interaction in the plane of the fan. In this case, the solution was to change the pylon geometry – its thickness and the shape of its leading edge. Other solutions include "staggering" the outlet guide vanes to even out the effects of the pylon pressure field so that there is a smooth (circumferential) pressure distribution in the plane of the fan.

A circumferential distortion in the inlet wall boundary layer in which the fan has to operate will also generate tones. Some high-bypass-turbofan engines have an intake system that is noncircular at entry, even visibly "flat-bottomed", to promote greater runway clearance. Under these conditions, there is a good chance that a tone will be generated as a result of circumferential inflow distortion.

Broadband noise is normally kept under control by minimising blade-wake effects, since its level is a function of velocity over the blade surface and the "activity level" in the blade wake. The detailed design of the blade, including the minimisation of aerodynamic losses, is crucial. This means that losses have to be addressed at what are really engine "off-design" conditions, to cover the operating powers

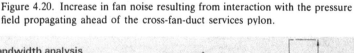

Figure 4.20. Increase in fan noise resulting from interaction with the pressure field propagating ahead of the cross-fan-duct services pylon.

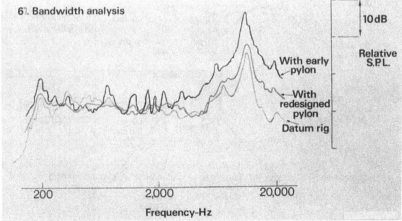

for take-off (full and cutback thrust) and landing, and not just at the cruise point, which is the main (aerodynamic) design point. For example, large fan-blade incidence excursions or movement of engine variables such as bleed valves and guide vanes should be avoided in the critical power range.

All the above considerations, as applied to the modern engine (see Fig. 4.21), have achieved noise reductions of some 20 PNdB when compared to the PW-JT3D, arguably the first "turbofan". Half of this improvement is due to the use of acoustically absorbent material, discussed in Section 4.3.

Moving to the rear of the engine, the turbine stages can produce as intense a noise as the fan and core compressor. This is partly because all the noise generated in the turbine stages is radiated rearwards, the sound field being unable to penetrate the nozzle guide-vane system behind the combustor, which runs choked and presents a "sonic" barrier. Moreover, turbine stages are usually much closer together than fan and early compressor stages, and this in itself leads to high levels of noise, owing to both pressure-field and wake interactions, It is often because the total massflow through the core engine is only 20–25% of that through the total fan system (and hence 6–10 dB quieter at the same surface velocity) that the turbine is not the major noise source on this type of engine.

Temperatures in the turbine section are, of course, much higher

Figure 4.21. Fan noise control – evolution of the quiet turbofan.

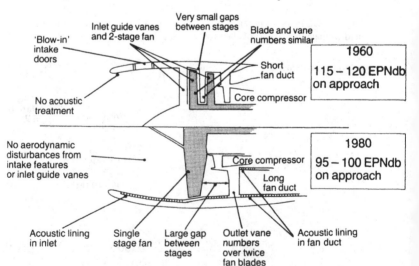

than in the fan and the compressor and, accordingly, the processes of noise generation are affected. For example, the ratio of stationary to rotating blades necessary to promote cut-off is less than in the fan, owing to the change in the speed of sound, and intensity is affected by density changes. Nevertheless, where interaction tones need to be suppressed, the same fundamental noise-control techniques apply in the turbine as in the fan. Spacing should be adequate so as to avoid pressure-field interactions if possible, and to minimise wake interactions, and numbers can be chosen to promote cut-off. However, because of the multiplicity of stages, it is difficult to avoid tones altogether, not to mention the additional "sum-and-difference" effects. To control noise in the turbine it is often necessary to use acoustic absorbers in the exhaust duct and, to a certain extent, to rely on "smearing" of the sharp tonal signature as it propagates through the turbulent exhaust-mixing process. This effect can often be enhanced by the additional turbulence generated if an internal forced mixer is utilised to accelerate the mixing of the hot core and cold fan-bypass flows.

Although noise generation in the combustor section is not yet fully understood, and no conscious action is known to have been taken to control noise generated here, it is a source that needs to be borne in mind. Fortunately, however, it is far less important to control this noise than that produced by the jet mixing process.

Jet noise can be minimised in one of two ways. Firstly, once the engine cycle has been determined, the geometry and relative positions of the primary and secondary nozzles can be optimised. For example, the fore and aft (relative) positions of the two nozzles, and the presence of a plug in the hot stream, can all influence the velocity profile downstream of the nozzle system and hence affect the generation of mixing noise.[824] No generalised noise-control rules have been developed, but it is well known that the detailed design features of a complex exhaust system are likely to affect the level of jet noise by ± 2 dB. The normal procedure is to investigate and optimise the system using scale models.

Secondly, as an alternative, the fan and core flows can be mixed upstream of a single final nozzle, to eliminate the influence of a high-velocity core flow. In recent years, some turbofan designs have come to rely on mixers of the type shown in Figure 4.22, as they reduce noise by up to 4 dB, in comparison with unmixed streams.

There are still other sources that often require attention. For example, the compressor bleed valves, which allow high-pressure air to be offloaded from the compressor system at critical points in the operating range, often inject a high-velocity jet of air directly into the bypass duct. Bleed-valve noise from a high-pressure compressor can be amongst the loudest individual engine noises at low power. Its suppression has reduced the overall aircraft noise on approach by as much as 3–4 EPNdB. One way to reduce bleed-valve noise is to introduce a pressure-reducing device followed by a multihole outlet (Fig. 4.23). The pressure-reducing stage lowers the final efflux velocity, whilst the multihole outlet subdivides the compressor offload flow into a number of small discrete jets. These induce rapid mixing and produce high-frequency noise that is then successfully attenuated by the acoustic lining in the bypass duct. In all, the process is analogous to that introduced by the jet suppressor and the lined ejector.

The acoustic lining is perhaps the most important "new" mechanical

Figure 4.22. An internal hot/cold flow-stream mixer.

feature of the modern engine. It needs to be discussed in some detail, for it contributes as much to noise control in the turbofan as any of the factors discussed already. (See Fig. 4.24.)

4.3 Acoustic absorbers

Of all the beneficial changes that have taken place in the design cycle of the aero gas-turbine in the past twenty-five years, half of those that have contributed to noise reduction involve the use of acoustically absorbent material in the air and gasflow ducting of the power-plant. Typically, absorbent liners in the intake of the power-plant account for about 5-PNdB noise reduction, whilst lining in the discharge ducts, both hot and cold, account for over twice this amount.

What accounts for the lining's "magic"? The answer, of course, is that there is no magic, for acoustic liners have been used commercially for many years, notably in buildings where it is necessary to improve the acoustics or reduce the general noise level in what would otherwise be a reverberant chamber. In the case of the jet engine, the important difference is that it does not have a very large surface area to which the linings may be applied and, of course, weight is a prime consideration. Equally, the environment in which linings have to survive is extremely unfriendly. Temperatures range from $-50\,°C$ to $+500\,°C$ between the limits of the intake air flow at high altitude and the hot gas temperatures in the exhaust system.

Even so, lining performance is good and can be predicted reasonably well. There is by no means the same uncertainty surrounding the effectiveness of acoustic treatment as there is in accurately predicting component noise levels in the turbomachine. Liner effectiveness is a function of several parameters; some have to do with the configuration of the liner itself, others with the environment in which it has to operate.[530-44]

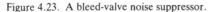

Figure 4.23. A bleed-valve noise suppressor.

Figure 4.24. Typical noise-control features of a turbofan.

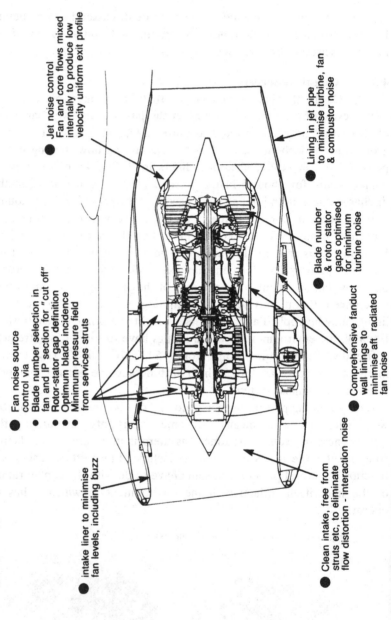

Fan noise source
control via
- Blade number selection in
  fan and IP section for "cut off"
- Rotor-stator gap definition
- Optimum blade incidence
- Minimum pressure field
  from services struts

Jet noise control
Fan and core flows mixed
internally to produce low
velocity uniform exit profile

Lining in jet pipe
to minimise turbine, fan
& combustor noise

Blade number
& rotor stator
gaps optimised
for minimum
turbine noise

Comprehensive fanduct
wall linings to
minimise aft radiated
fan noise

intake liner to minimise
fan levels, including buzz

Clean intake, free from
struts etc, to eliminate
flow distortion - interaction noise

Because of the hostile environment in which liners have to operate, there is an obvious need for structural rigidity. This requirement runs counter to the more overriding consideration of power-plant weight and, as a result, the lining structure is often configured so that it can strengthen the nacelle in which it is installed. For this reason, the majority of liners in service throughout the wide range of power-plants that employ them comprise a simple perforated-plate face sheet supported by a honeycomb structure a specified distance from the solid wall of the engine ducting. Although by no means the optimum acoustic design, this form of lining has been calculated to provide 3–5 EPNdB of the total attenuation possible if it was replaced by liners with "perfect" absorption at all frequencies. This, indeed, is a commendable achievement, when one bears in mind the mechanical demands placed on the structure, including the possibility of destruction due to freeze–thaw cycling of entrained water or the danger of fire resulting from entrapped fuel and oil.

As Figure 4.25 shows, acoustic linings that have been considered for inclusion in aircraft power-plants range from simple commercial fibrous or "bulk" materials to complex multiple-layer assemblies. In all these designs, two basic properties are being exploited, albeit to different degrees: (1) the physical damping of the pressure fluctuations in the porous "resistive" structure of the face sheet, and (2) the "reactive" cancellation of the direct incident sound-wave by the wave that is reflected from the solid backing structure inside the liner (see Fig.

Figure 4.25. Possible acoustic liner configurations.

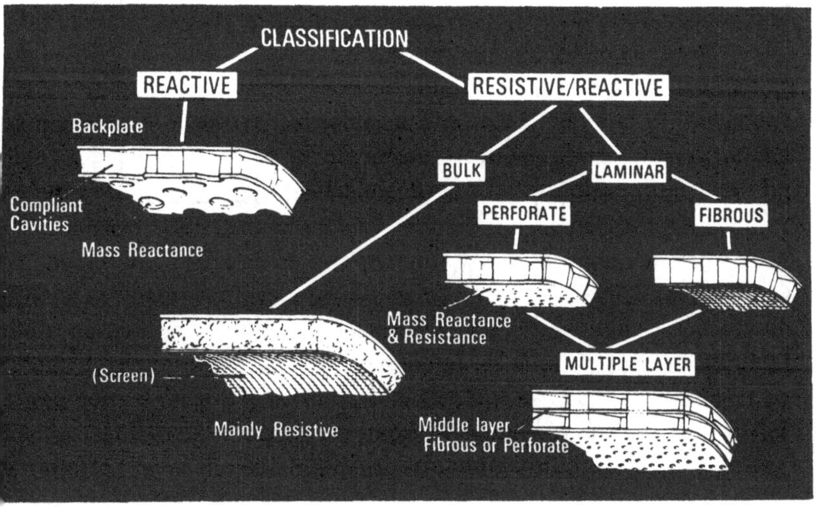

Figure 4.26. Mechanism of noise absorption by an acoustic liner.

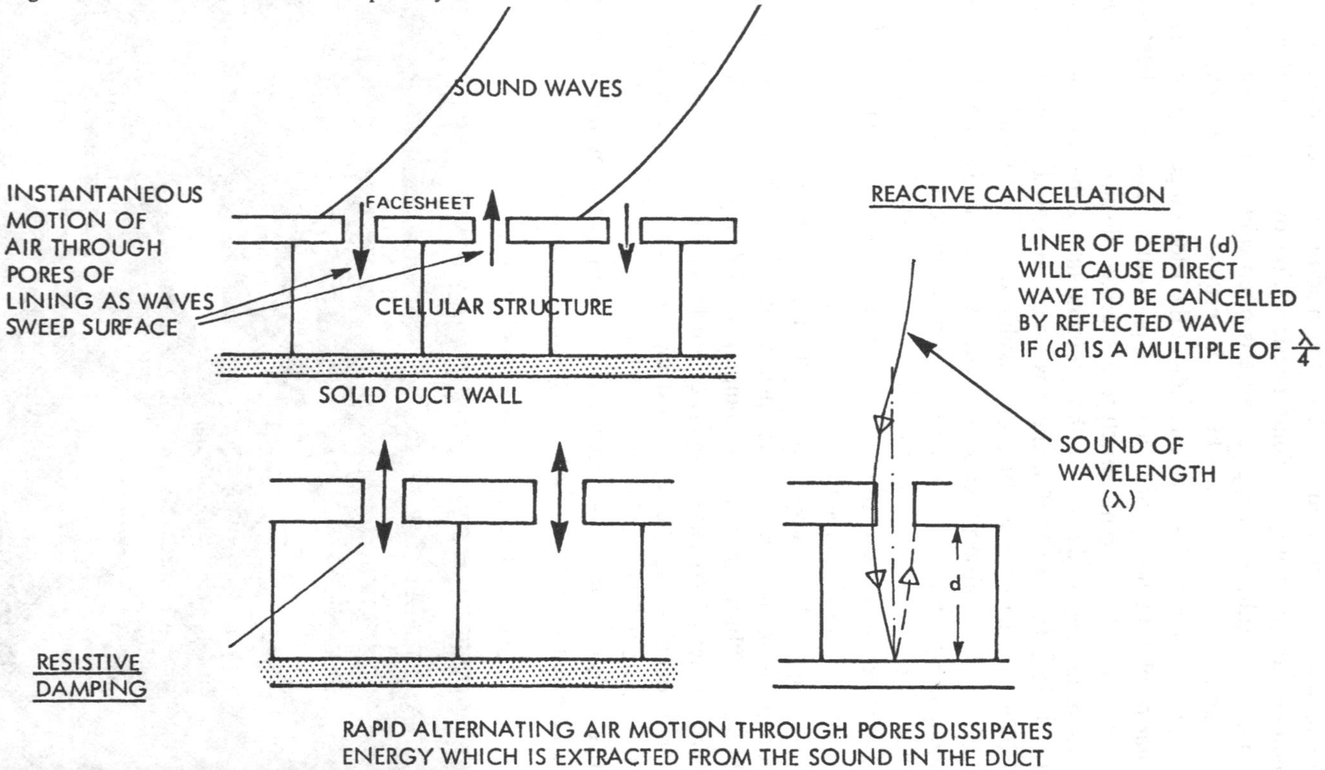

REACTIVE CANCELLATION

LINER OF DEPTH (d)
WILL CAUSE DIRECT
WAVE TO BE CANCELLED
BY REFLECTED WAVE
IF (d) IS A MULTIPLE OF $\frac{\lambda}{4}$

SOUND OF
WAVELENGTH
(λ)

d

SOUND WAVES

INSTANTANEOUS
MOTION OF
AIR THROUGH
PORES OF
LINING AS WAVES
SWEEP SURFACE

FACESHEET

CELLULAR STRUCTURE

SOLID DUCT WALL

RESISTIVE
DAMPING

RAPID ALTERNATING AIR MOTION THROUGH PORES DISSIPATES
ENERGY WHICH IS EXTRACTED FROM THE SOUND IN THE DUCT

4.26). In the simplest case, the one-hole-per-cell perforated-plate liner, the wavefront is presented with what is essentially an array of tuned resonators, with the distance between the face sheet and the solid backing structure being the mechanism for varying the tuning frequency, or frequency of peak absorption. There is a small additional effect that is related to the inertia of the air or gas moving in and out of the face-sheet pores and to the compressibility of the gas in the cavity.

The reactive properties of a liner can only be exploited fully if it can be guaranteed that there are narrowband sources of noise that do not vary in frequency with the operating conditions of the engine. It will be readily appreciated that this is not the normal situation in an aircraft engine, where the important operating conditions (from the noise standpoint) occur at high power during take-off and at low power during the landing approach. Consequently, the damping property of an acoustic liner has to be put to good advantage to ensure worthwhile sound attenuation over a wide range of frequencies.

The resistive damping effect is inversely proportional to the porosity of the face sheet. As a result, typical aeroengine liners have a low porosity (about 5%), which creates manufacturing problems. It is difficult to permanently bond the honeycomb supporting-structure to the face sheet without locally reducing the porosity, by blocking some of the pores with bonding material. It is important to prevent such blockage in order to ensure that the porosity of the face sheet will remain uniform and that peak resistance will be achieved. The gross porosity of facing material can be determined approximately by measuring the pressure drop through the material with a steady flow of air. However, resistance does vary nonlinearly at high flow rates, and allowances must be made for this at very high sound-pressure levels. Resistance also changes with a flow of air over the surface; in fact, it increases markedly with increasing flow Mach number.

Because resistance cannot be directly obtained from simple laboratory bench testing, it has been necessary to develop correlations between determinable laboratory parameters and the effectiveness of acoustic linings under full-scale engine conditions. Although attempts have been made to measure impedance in situ, they are only just becoming a common technique.

The variation of attenuation with resistance is illustrated in Figure 4.27, which also catalogues three other important effects: the airflow direction in the duct (with respect to the sound propagation path), the absolute length of the duct lining and the ratio of liner length to duct depth. Also shown is the tuning effect of liner internal depth. The

Figure 4.27. Parameters affecting the performance of an acoustic liner.

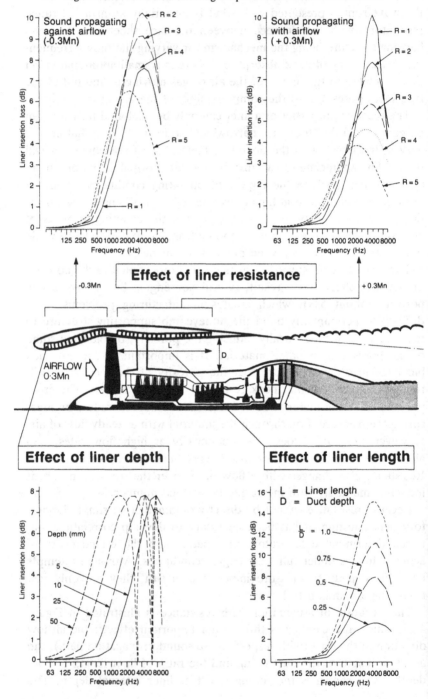

importance of the ratio of duct length to depth (L/D) is simply but graphically illustrated in Figure 4.28 where, in terms of simple ray progression, a large value of L/D gives the opportunity for multiple wall reflection of high-order modes and hence greater chance for a liner to attenuate the pressure fluctuations. This is why fan-duct linings are usually more effective than those in the air inlet, where duct "depth" is the whole diameter, often over 2 m, and length is as short as the efficient aerodynamic handling of inlet airflow will allow.

Generally speaking, the most effective liner is the type with a "bulk" configuration. Commercially, it takes the form of interconnected-cell foamed plastic or a fibrous material and is used in applications ranging from anechoic chambers to industrial and domestic ceilings. A liner of the fibrous form will be faced with a high-porosity screen to hold the fibres in place. This may be a woven sheet or a perforated plate, but, because of its high porosity, it allows the material it contains to exploit one of its least-desirable properties – a propensity for fluid absorption and retention. This property has meant that bulk absorbers have not yet found application in the jet engine – either in the cold ducts (where freeze/thaw action on retained water is the problem) or in the hot ducts (where combustible-fluid retention is a fire hazard). As a consequence, the quest continues for cheap, lightweight approximations of the bulk absorber that can be applied in the limited space available in the engine without the hazards of the "real thing". Some of the structures used in service to date are indicated in Figure 4.29. The

Figure 4.28. Opportunities for noise reduction by acoustic lining. (From Ref. 211, by permission)

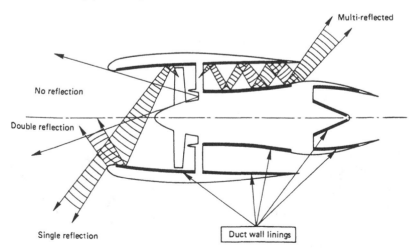

Multi-reflected

No reflection

Double reflection

Single reflection

Duct wall linings

configuration with a woven face sheet, the so-called linear liner, has better bandwidth of attenuation than the simple porous face sheet, and it maintains its properties over a wider range of environmental conditions in the ducting. The multiple-layer approach improves on these properties further by presenting the sound-wave with various internal depths and layers of resistance. Although such constructions are more difficult to manufacture and heavier than single-layer designs, they are being used more and more, particularly in the air inlet, where the

Figure 4.29. Typical in-service acoustic liner constructions.

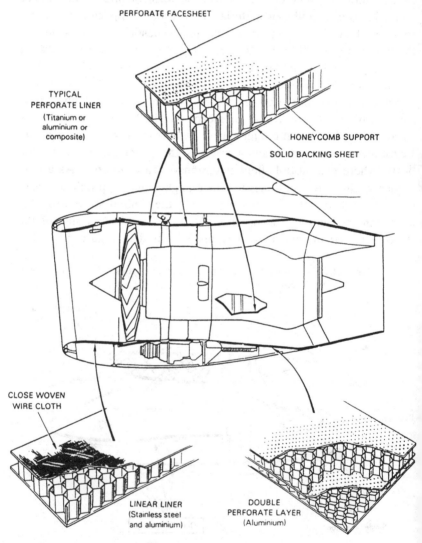

PERFORATE FACESHEET

TYPICAL
PERFORATE LINER
(Titanium or
aluminium or
composite)

HONEYCOMB SUPPORT

SOLID BACKING SHEET

CLOSE WOVEN
WIRE CLOTH

LINEAR LINER
(Stainless steel
and aluminium)

DOUBLE
PERFORATE LAYER
(Aluminium)

physical duct dimensions go counter to good sound attenuation configurations and where every $\frac{1}{10}$ dB in sound attenuation is worth achieving.

To be effective in controlling the overall noise of the engine, the wall liners have to be tailored to the source characteristics and the duct conditions. The means by which sound propagates along engine ducts are complex.[500-13] Consequently it is not yet clear which liner design is most appropriate. Broadly speaking, the design process involves varying the internal depth of the liner (to control peak frequency of absorption) and its detailed structure and the face-sheet porosity (to control bandwidth). The high frequencies generated in the turbine lead to very thin liner structures, of the order of a centimetre, whereas low-frequency fan "buzz" tones may demand a depth of several centimetres. In the modern engine, the judicious use of liners accounts for half the noise reduction that the turbofan shows over earlier engine cycles, typically 10 PNdB.

Figure 4.30 illustrates this point by showing how a simple single-layer perforated-plate face-sheet construction reduces the noise radiating from the rear of a turbofan by an average of 10 dB over the frequency range 1–11 kHz, as a result of attacking the fan and core engine sources in the two separate exhaust ducts. Further improvements result from the advanced structures discussed earlier, and by increasing the surface area available for the application of liners, such as splitters in the ducting. Care has to be exercised to avoid creating

Figure 4.30. Narrowband analysis showing the effect of fan-duct and core-engine duct liners.

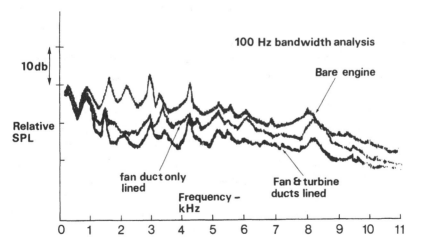

new sources of noise, either by pressure-field or wake interactions with the rotating machinery, or from liner surface-induced broadband noise.[780-2] The value of increasing liner surface area has been demonstrated at full scale,[851] but the commercial application of the findings has to be weighed against the performance and weight penalties that can be incurred.[868-9]

4.4   Other noise control suggestions

Humanity is naturally inventive. The capacity for solving one's neighbour's problems shows no bounds, and the aircraft noise issue is no exception. The industry regularly receives offers of help, ranging from novel suggestions for noise suppression, like those expressed in Figure 4.31, to offers to sell "vital" patent rights. Most of the ideas are well intentioned but many are just out of date or lack an understanding of the problem, and it is often difficult to refute the suggestions without causing offence.

Nevertheless, between the amateur's suggestion and the reality of the commercial noise suppressor lies a battlefield of fallen ideas and research hardware. Some, like the ejector–suppressor previously discussed, have taken the form of devices that attack the noise after it has been generated. Others, like inlet flow control,[855-60] even to the point of "choking", try to keep the sound within the engine, whilst the ultimate approach is to remove the cause of the noise source.[853-4,861-3]

In this respect, observation of what happens in nature is sometimes the inspiration. For example, the owl is reputed to fly "silently" and its serrated wing edge form has been emulated in fan testing.[864-7] Nevertheless, fan blades still have uniform edge profiles and are far from silent. Could it be, perhaps, that the fact that the owl's wing is a large and lightly loaded device, displaying variable geometry, has some bearing on the issue? Or is it because the owl's terminal speed is, typically, only one-fiftieth of that of the tip of a large engine fan assembly which, at full power, is producing enough thrust to support over half a million owls?

Another idea, to skew the stator-row[853-4] with respect to the rotors, to "smooth" the blade-wake interactive-tone generation process, also fell by the wayside. In this case, researchers failed to recognise that the wakes shed by the stators were already skewed with respect to the rotors through the normal processes of swirling three-dimensional flow. In any event, although researchers did not know it at the time, they were probably studying the interaction of atmospheric turbulence with the rotor, and not blade-wake effects. Most fan and compressor

Figure 4.31. Expert advice from the public.

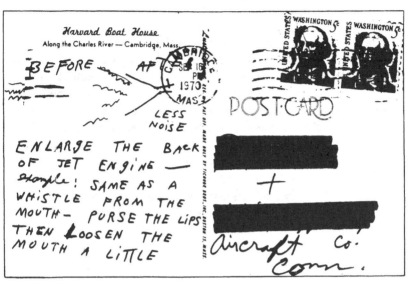

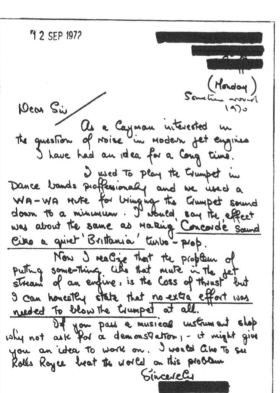

tests of the 1960s and early 1970s were confused by inlet airflow distortion-interaction effects, as were the early turbofan engine tests on open-air facilities (see Section 3.5).

Many serious tests were conducted in the interest of suppressing jet noise without the knowledge that other sources were confusing the issue. Model jet rigs were often excellent sources of combustion noise and disguised the actual results of jet mixing noise tests. Conversely, many jet suppressors that performed well at model scale on component test facilities did not have a worthwhile effect when flown on test aircraft,[836] owing to the effects of forward motion and the presence of other noise sources in the whole engine.

Examples of well-intentioned work foundering on nature's rocks abound, but they must not deter the experimentalist. Without the 90% or so of wasted tests, the fruitful few would not have provided the advances in aircraft noise control that we have seen since 1960.

5

Concorde – a special case

Conceived as the transport of the future, in an era when fuel accounted for less than 10% of airline costs, the Concorde (Fig. 5.1) was a special case by any yardstick. Capable of flying at Mach 2, or over 2000 kph, for distances of about 6500 km without in-flight refuelling, it outpaces most and outsustains all the world's military aircraft. In fact, in mid-1987, by which time the Concorde had carried its millionth passenger, it had also operated at supersonic speeds for more than twice the time of the whole of the Western world's military fleet put together.

Its commercial failure – only sixteen production-standard aircraft were manufactured – was entirely a function of the change in the world economy in the 1970s. With the price of aviation fuel rising almost tenfold in a decade, there was a sustained drive to reduce the fuel consumption of all aircraft engines. Accordingly, concerted efforts to alter the design of subsonic engine components, allied to the emergence of the turbofan engine, meant that the Concorde's subsonic contemporaries were achieving up to 40% better fuel consumption than their forebears. The Middle East war of the early 1970s, which sparked the alarming increases in the price of oil shown in Figure 5.2, only served to emphasise the Concorde's uncompetitiveness, for the 40% improvement in fuel consumption only softened the blow to the airlines, which still found fuel to be 30% of their overall operating costs and would have found the Concorde even more consumptive. Over half the Concorde's weight is taken up by fuel. Hence, it is hardly surprising that the production run was limited.

Nevertheless, the mere fact that it entered service was a technological achievement that many believed the combined forces of the United Kingdom and France could not deliver. However, in proving that they could, they also created an unsurpassed environmental furore. Apart from the debate on the Concorde's effect on the ozone layer and the world's climate in general, major concern centred around its noise problem. This problem was a direct result of the design of the aeroplane: It was designed to cruise at Mach 2, and therefore the exhaust velocity from the engines had to be sufficiently in excess of Mach 2 to overcome airframe drag and hence maintain stable flight. This called for a jet exhaust velocity and an overall engine-pressure ratio well in excess of those of the early pure jet engines.

Since it was natural to select a derivative of a military supersonic reheated pure or "straight" turbojet-engine cycle, there was never going to be a simple "solution" to the noise problem that would not have eroded dramatically the Concorde's payload-range capability. Nevertheless, every effort was made to achieve noise suppression within the known performance allowances although, when it entered service, special noise-control devices were not a feature of the Concorde's design. Despite these efforts, serious environmental concerns were voiced by the public, priests and poets. The day the Concorde first landed at Kennedy Airport the local protesters made so much noise

Figure 5.1. The Concorde.

themselves that they were unaware that the aircraft had landed! Since then, noise control has been exercised via carefully planned and unstintingly executed flight manoeuvres that have ensured trouble-free operations.

The cost of placating the public was not insignificant. From day one, the project planners recognised that there was a noise concern, and so they instituted a basic design goal of levels "no noisier than existing long-range subsonic aircraft". Unfortunately, in the time period between the Concorde's conception in the late 1950s and its first fare-paying passenger service almost twenty years later, "existing" subsonic jet aircraft had changed markedly – not only in shape and size, but in the noise they produced, which could be anything up to 20 dB less than the DC8, B707, CV880 and VC10, which were the Concorde's contemporaries during its early development phase. The whole environmental backdrop to the Concorde story is interesting in its own right.

5.1 Chronology of events

As is already extremely well documented,[17] the early meetings of the U.K. Supersonic Transport Aircraft Committee (STAC) recognised that the engines selected for an SST would generate more noise than that emitted by contemporary jet engines. Nonetheless, it was believed that methods of suppression would be found that would ensure environmental compatability. An extract from the STAC main report in 1959 carried this message:

> An existing problem, that of aircraft noise on the ground, is much accentuated with engines designed for optimum per-

Figure 5.2. Variation with price of crude oil over the course of this century.

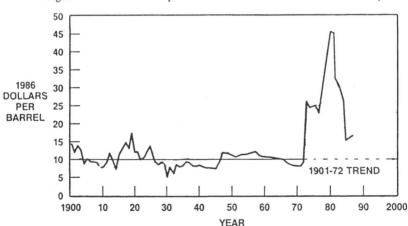

formance at supersonic speeds and the problem of moderating such noise without incurring prohibitive penalties is crucial to the whole project. Whereas the noise of subsonic jets could be reduced without performance penalties by using special engines with low jet velocities, this technique is less and less applicable as one goes up to the supersonic speed range. A wide range of methods for throttling techniques in the climb and engine silencing devices has therefore been considered. ...

The noise levels under the climb can be reduced considerably by throttling the engine soon after take-off ... the supersonic aircraft will benefit especially from such techniques because the fully variable exit nozzle ... allows a reduction in thrust without reducing the air mass flow. ...

The final answer will undoubtedly depend critically on the detailed design of the aircraft and the engine installation but the Committee believes that there is a fair prospect of finding solutions without excessive penalties and that the noise problem should be most carefully studied in the design studies that are recommended.

Perhaps it would have been more accurate to substitute the word "hopes" for the word "believes" in the last sentence. Despite the honourable intentions expressed, it could be claimed that the designers were guilty of "travelling hopefully" until it became obvious that even the noise levels of the existing jet fleet were wholly unacceptable to the public in general. The events of 1966, the London Noise Conference and the FAA letter to U.S. industry proposing the institution of noise certification, drove home to the Concorde consortium the need for positive action. Faced with the possibility that any new noise certification requirements to emerge from international discussions might touch upon the commercial viability of the Concorde, the U.K. and French partners involved in the development and delivery of the power-plants (Rolls-Royce and SNECMA) were driven to institute noise R&D programmes at a level significantly higher than anything previously witnessed. A special Jet Noise Panel[18] was instituted, which brought together the technical expertise in the French and U.K. industries, and the assistance of eminent researchers from universities and government establishments. Although its objective was primarily to seek a long-term "solution" to the general problem of high-velocity jet-exhaust noise, it did oversee a highly concentrated experimental programme directed at finding an engineering solution to the noise problem created by the four reheated Olympus jets that were to take the Concorde into service.

Interestingly, whilst this was happening, an equivalent upsurge in activity was taking place on the subsonic jet front; in the context of the novel high-bypass-ratio turbofan. The number of scientists and engineers working on each of these issues in the United Kingdom alone rose tenfold, and expensive new specialised facilities were built with considerable government financial and technical support. A purpose-built "front-end" (fan and compressor) noise-research facility was established at the geographic centre of the industry in Derby and Bristol and at the government's National Gas Turbine Research Establishment at Pyestock, near Farnborough. New model-scale jet facilities were commissioned both at Pyestock and Bristol, and new full-scale engine test-beds at Aston Down (close to Bristol) and Hucknall (close to Derby). The tools for research into the effectiveness of sound-suppression linings within the inlet duct were provided in more facilities at both Hucknall and Pyestock, and eventually all turbine- and jet-noise facilities were replaced by a second major capital investment at Pyestock, the National Noise Test Facility. Funding for these facilities, and for intensive research activities, was provided from a special environmental "purse" administered by U.K. government departments. Similarly, in the United States, the National Aeronautics and Space Administration (NASA) and the Department of Transportation (DOT) were letting valuable contracts to all who had the capability of conducting noise research.

The results of all these efforts are now well known. In the context of the subsonic high-bypass-turbofan engine, they were extremely successful; in the context of the Concorde, they were close to a total failure. Although automated nozzle-area control, an aerodynamic-performance necessity on the power-plant, was used to good effect to reduce jet velocity at low powers, almost a decade of experimental and theoretical activity was finally buried when the sole surviving suppression devices in the Concorde power-plants were removed late in the development programme because they were ineffective; this move was referred to in the press as a "weight-saving measure".

Well before this happened, a debate was raging as to which countries would accept the Concorde at their airports, and which ones would permit the aircraft to overfly their territory and allow their citizens to be jolted by the occasional sonic boom. Anti-Concorde sentiments were most vehement in the United States. That is not to say that environmentalists in Europe, particularly in the United Kingdom, did not voice their opinion. In fact, they lobbied actively with the anti-Concorde groups in the United States, using noisy and fuel-greedy first-generation jet aircraft to carry them to their conventions, where they themselves fueled the flame-throwers that were jeopardising the initial use of the first sixteen Concordes, let alone any further orders.

In several respects, the anti-Concorde lobby was extremely successful. One of the Concorde's marketing platforms was its ability to make the transatlantic trip several times a day, in the limit allowing VIPs to make a swift "day trip" to the East Coast of the United States. To do this, the aircraft had to be capable of both taking off and landing at major U.S. and European airports at night. This cornerstone was hacked from beneath the structure of the Concorde programme early on. In fact, even its operation in the daytime became a prime concern of the manufacturers and of Air France and British Airways, who were to become its only owners. The situation in the United States was the major hurdle. By 1969, U.S. citizens had become so environmentally conscious that any major federal action likely to affect the quality of life there had to be subjected to open scrutiny. The potential acceptance of Concorde operations at U.S. airports fell nicely into this category, and the most exciting aircraft project for years became the aviation guinea-pig for an emerging environmental examination process. The formal letters from British Airways and Air France, requesting permission to operate the Concorde in the United States, set in motion an official chain of events that included the preparation of one particularly important public document, an environmental impact statement (EIS).

The EIS would be a significant, although not overriding determining factor in any decision to allow the Concorde to operate in the United States. Ultimately, the decision lay with the secretary of transportation, who had to weigh not only the environmental issues but the commercial and international implications. The EIS was prepared by his aviation department, the Federal Aviation Administration (FAA), and bore heavily upon contributions made by the British and French governments, the aeroplane manufacturers (Aerospatiale and British Aircraft Corporation) and the joint power-plant manufacturers. The FAA published a draft EIS in March 1975 and followed this with public hearings in Washington and New York, the principal targets for Concorde operations. These were followed by a final EIS[165] in November 1975 and a public hearing on 5 January 1976 presided over by the secretary of transportation himself. Following this public hearing, the secretary made his historic decision[19] to allow the Concorde to fly into the United States for a trial period of up to sixteen months. The decision, published on 4 February 1976, allowed up to two Concorde flights each day by both Air France and British Airways into John F. Kennedy Airport, New York, and one a day by each airline into Dulles Airport, just outside Washington, D.C. These flights had to be made between the hours of 7:00 a.m. and 10:00 p.m. In addition, the Concorde was prohibited from making supersonic flights, and hence sonic booms, over the continental United States.

The secretary's decision reflected the height of diplomacy for, in terms of community noise, the Concorde represented a step backwards as much as the contemporary turbofan engine represented a step forwards. Even so, opposition to the Concorde was not limited to the noise issue, for both the Environmental Protection Agency (EPA) and the Federal Energy Administration (FEA) claimed adverse energy and other environmental impacts. Moreover, on the commercial front, even the U.S. carriers Pan American and Trans World Airlines opposed it on the basis that it would cream off their first-class passenger traffic, despite the fact that Pan American had signed Concorde's first purchase option in 1963, and subsequently cooperated with the British and French flag carriers to lay down basic production specifications.

Although the transportation secretary's decision was taken on 4 February 1976, and operations into Washington were inaugurated 110 days later, the Concorde saga dragged on, so that it took almost two years before it was able to operate from New York's Kennedy Airport. Although there were local legal objections to the Dulles operation, the FAA owned the airport and were able to prepare rebuttals quickly, in the cause of implementing their boss's decision. Hence there was little delay before Air France and British Airways started the Concorde's North Atlantic services from Paris and London. Kennedy Airport was quite a different story, not being federally owned but administered by the Port Authority of New York and New Jersey, which regarded any final operational decision as theirs and not subject to any federal edict.

Kennedy is a busy airport – far busier than Dulles, which was deliberately built 25 miles from the centre of Washington, D.C. to move the noise problem (and some traffic) out of town. During the 1970s, it was shunned by the airlines whenever possible, and by passengers who wished to spend as little of their journey time as possible commuting between the city and the airport. Kennedy was a busier and a noisier airport, and with the population density surrounding it being much higher, people readily expressed their views directly to the Port Authority and through the courts. Some half a million people then resided within the boundaries of the critical 30-NEF (noise exposure forecast) contour around Kennedy, whereas less than a thousand could describe themselves as similarly "impacted" by noise from Dulles. As a result, and hardly surprisingly, the battle-scarred Port Authority took the initiative over the question of Concorde operations and, unilaterally, decided to ignore the FAA and ban its entry. Immediately, both British Airways and Air France filed lawsuits against the Port Authority, and a protracted and tardy legal process was instigated. The initial British Airways–Air France suit found favour in the Federal District Court, but subsequently that decision was overruled on appeal by the Port Authority. In a second airlines' petition, the Port Authority was ruled to be discriminatory and ordered to allow the limited number of operations decreed by the secretary of transportation. This decision was upheld despite a further appeal by the Port Authority, which even tried to refer the matter to the U.S. Supreme Court, but failed to gain the necessary interest and support.

The Concorde finally landed in New York on 19 October 1977 and delivered fare-paying passengers five weeks later, on 22 November, twenty-two months after it started operations from London and Paris to Bahrein, Dakar and Rio de Janeiro. Thanks to the experience gained at the Europearn airports, and all the careful noise-abatement planning and practising, the first arrival at Kennedy was well within the monitored noise limits and went unnoticed by the bands of protesters who gathered to cause disruption. Equal care was taken in planning the subsequent take-off operation, with low-altitude power reduction and tight turns keeping the noise down in the most sensitive communities around the airport.

From then on, with aircraft operating into both Washington and New York, it was possible to evaluate the impact they had on the noise climate around the two airports.[32] The noise level at Dulles proved to be consistent with predictions and, although complaints were received, the public gradually began to accept the Concorde's presence. Eventually, at the end of the trial period, the Concorde was granted permission to land at thirteen U.S. airports. Subsequently, although Concordes with flight time before 1980 were exempt from any further restrictions, to ensure that a similar situation never arose again, the FAA introduced into its regulations both noise and sonic boom requirements for all future SSTs.

Sonic boom was a sensitive issue during the period of Concorde flight testing and during the early days of service. The "overpressure" of the boom, or aerodynamic bow wave, that the passage of a vehicle at speeds greater than sound causes to be propagated to the ground, produces sounds akin to a sharp whipcrack, thunder, gunfire or an explosion. As explained in Section 3.2, the intensity of the sonic boom is a function of the shape and size of the aircraft, its speed and its altitude. Moreover, some climatic conditions can induce secondary and tertiary booms as a result of atmospheric refractive and reflective effects.

Sonic boom has been the subject of research both in Europe and the United States for a considerable period of time. It has involved theoretical work, laboratory experiments and aircraft tests. These have utilised a range of military aircraft and the Concorde itself. During development flying in the 1970s, a north–south corridor to the west of the United Kingdom became labelled "bang alley" and great concern was expressed about possible structural damage to ancient buildings in the extreme west of both Wales and Cornwall.

Fortunately, measurements in a sample of ancient buildings showed that there should be less concern about the Concorde's operations than about a thunderstorm, and that buildings in major cities were subject to vibrations many orders of magnitude greater as a result of day-to-day traffic.[733-4] Nevertheless, the British government adopted a liberal attitude and compensated, almost without question, any damage claimed by inhabitants of the southwest of England and Wales. Subsequently, when the Concorde was in regular service, the sonic boom did prove to be a concern during the approach to the major airports in France, the United Kingdom and the United States.[32] Typically, the United States tackled the issue through legislation, prohibiting the generation of sonic booms over land. The British and French governments, the sponsors of the Concorde project, were more relaxed in their attitudes. They tended to allow the airlines to sort the problem out via route alterations and other operational adjustments that addressed the areas of complaint, typically the southern coast of Britain and the northern coast of France.

5.2 What are the lessons of the Concorde?

Purely in terms of the general noise issue, the Concorde teaches an important lesson – that no manufacturer should ever again even contemplate producing an aircraft whose noise levels are not compatible with those generated by the best contemporary aircraft.

The Concorde's enthusiasts "travelled hopefully", thinking that if Concorde were no noisier than the (then) noisiest contemporary aircraft there would be no problem. Some also believed that nobody would stand in the way of a project that was extending the frontiers of knowledge. The fact that there are only a handful of Concordes in service is probably why the aircraft has been able to continue operating, for it is exceedingly noisy.

Another obvious lesson is that, if the noise targets set by the best contemporary aircraft are not immediately achievable by the engine design cycle in combination with the aerodynamic performance of a new aircraft, then the technology necessary to achieve the design goals should be proven before the project is launched. In the case of the Concorde, when it became apparent that the noise issue was not going to be "all right on the night" and that things were falling foul of the tremendous improvement brought about by the high-bypass turbofan, even a "fire-brigade" research and development programme was unable to cope with the belated and unreal target that was set.

Between the launching of the Concorde project and its entry into service, subsonic propulsion technology had moved forward dramatically. The high-bypass turbofan had appeared towards the end of the 1960s and, by the time the Concorde entered service, four major new aircraft had demonstrated to the public the enormous noise improvements resulting from the design cycle of such an engine. Consequently, far from being almost compatible with the "noisiest contemporary" aircraft, the Concorde suddenly stuck out like an environmental sore thumb. The final noise target, set late in the programme, was to reduce the Concorde's average noise level by some 10 dB, but this was clearly impossible to achieve. The wide-ranging model and full-scale research and development programmes that resulted, overseen by the power-plant manufacturers' special noise panel of experts, represented a last-minute bid to salvage a worsening situation. Although they failed in their overall objective, they did raise some interesting technical issues.

5.3   Technical issues

Almost all noise-related activities centred around the extremely high jet velocities that the supersonic engine cycle demands. The Concorde was designed to travel at a speed that corresponds to just over twice the speed of sound, or a Mach number just greater than 2. Subsonic aircraft cruise at less than half this value. It follows that the exhaust velocity at cruise needs to be high, by definition, sufficiently in excess of the cruise speed to account for the steady drag of the aeroplane at that condition. Early pure jets often had exhaust velocities in excess of 600 m per second, well in excess of the velocity necessary to sustain flight, but necessary because of the small mass of air involved in the process of generating momentum. Today's high-bypass engine merely exchanges velocity for massflow but there is a limit to the process – the momentum equation does not allow you to drop exhaust velocities below the speed at which you wish to fly!

With the added problem of having to control airflow into the engine whilst cruising at speeds in excess of twice the speed of sound, the optimum cycle for the Concorde's engines tends towards low mass-flows, or the selection of the straight turbojet. Moreover, because of size constraints and in order to provide adequate take-off performance and permit the aircraft to accelerate whilst climbing to an 11-mile-high cruising altitude, it becomes necessary to augment the thrust periodically by "reheating" or "afterburning" the exhaust flow for short periods. In this process, an additional fuel supply to the jetpipe behind the turbine allows unburned oxygen to be utilised in a further heat-release process that increases the energy in the jet. It is a common feature of military strike aircraft and, in fact, the Olympus engine that fulfilled all the performance requirements was a derivative of the engine selected to power the BAC-TSR2, an aircraft that was designed to provide the United Kingdom's supersonic tactical nuclear strike capability. However, during its flight development programme, it became the target of a newly elected socialist government's economy axe at a time when the Concorde project itself hung in the balance.

With jet velocities more than 50% higher than on the early pure jets, the flow from the exhaust is even more "active". The turbulent eddies that effect the mixing process with the atmosphere are, at low aircraft speeds, travelling at speeds (relative to the atmosphere) in excess of the speed of sound. They are in effect the generators of a multitude of sonic "boomlets". We can model this mixing process on a water table where, as Figure 5.3 shows, radiating sound-waves are visible. It is these Mach waves, coupled with an oscillatory motion of the jet plume (just visible here and in Fig. 5.5) and noise sources associated with shock cells that produces the violent crackling and tearing sounds that characterise many fighter aircraft and, of course, the Concorde itself on take-off.

Unfortunately, there is not much hope of suppressing the high levels of jet noise without prejudicing the solutions to complex aerodynamic problems. The need to provide systems to handle both intake and exhaust flows at low-speed/high-flow conditions on take-off, with efficiency and economy at cruise and low-speed/low-flow conditions on descent from cruise, places heavy demands on the technologist and the designer. It becomes necessary to provide a system that will make it possible to vary both the intake and the exhaust areas and, at the same time, provide the optimum intake and exhaust flow profiles; for example, there is a need to counter the losses and shock noise resulting from the sudden expansion of a supercritical jet. Couple this issue with that of a requirement to reverse the direction of thrust during the decelerating process after the aircraft has landed, and we find that the four Olympus engines that perform this wide range of duties required extremely innovative variable-geometry intake and exhaust nozzle assemblies. The exhaust is illustrated in the top half of Figure 5.4. The interesting feature of this assembly is the secondary nozzle, which performs several important functions between the start of an operation and its completion after landing. It is a fully variable device, taking the form of two "clam-shell" doors, or "buckets", as they are known colloquially. These can either squeeze the jet (and hence alter the flow geometry at the nozzle exit) or close completely to reverse engine thrust (or, more accurately, direct the exhaust slightly forwards rather than wholly rearwards). The main operating modes of the buckets are shown in the lower half of Figure 5.4.

Figure 5.3. A shallow water-table simulation of the noise field generated by a supersonic jet.

It was the ability to modify the shape of the jet by slightly closing the bucket assembly that led to an interesting observation and the bonus of some modest noise reduction during take-off, when the engines are at full reheated power and at their noisiest. It was noticed that, if the buckets were slightly closed into the form of a notch, the exhaust flow from the primary nozzle became squeezed in a horizontal direction and the combined flow from the two adjacent engines on each wing became slightly "fishtailed". Figure 5.5, a shadowgraph technique of flow visualisation, gives an impression of this squeezing effect with a simple "notched" nozzle. Although the resultant flow asymmetry had little effect on thrust, it caused a slight, but worthwhile lowering of the noise radiated to the side of the aircraft – some 2–3 dB in the Concorde's case.

Much scale-model noise and aerodynamic experimental work was

Figure 5.4. Concorde – Olympus 593 Type 28 exhaust system and main operational modes.

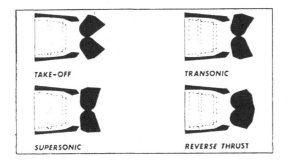

undertaken on notched-nozzle assemblies. The experimental results at
model and full scale correlated well and, by the time the Concorde
entered service, the optimum position for the buckets during take-off
and initial climb phases had been established, as shown at the bottom
of Figure 5.4.

The "bucket effect" was an unexpected bonus, and fortunate in that
it was the ground roll and initial climb sector of operation that caused
a major concern during the development programme, for it is one of
the operational conditions where the distance between the aircraft and
the community is fixed. Under the take-off flight path, it is always
possible to improve the performance of the aircraft so as to gain

Figure 5.5. Shadowgraph pictures of the "fishtail" jet from a notched nozzle.

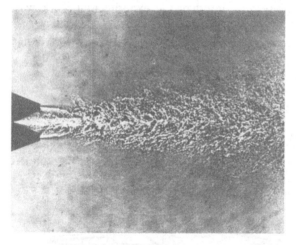

PLANE OF NOTCHES
(QUIET PLANE)

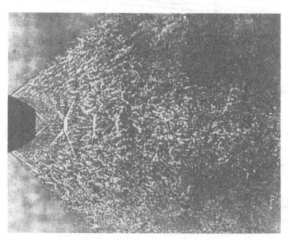

NORMAL TO PLANE
OF NOTCHES

altitude and hence put distance between the source and the receiver. Alternatively, it is possible to reduce power below maximum and lower the noise quite appreciably when overflying critical communities. In the event, use of the thrust-reverser buckets to slightly modify the jet mixing process was the only noise-reducing option possible at maximum power. At lower powers, for noise-abatement climb-out and on approach, the fully-variable nature of the nozzles allowed the control system to be programmed to ensure the highest massflow possible at a given thrust level. This produced the lowest jet velocity and minimum jet mixing noise. Otherwise, noise control was exercised by optimising aircraft speed, configuration and engine power to suit the operational needs at each airport. All this, despite the fact that the Concorde's power-plants ended their development life with a series of "spades" fixed just upstream of the final nozzle. The intent of these devices was to disturb the regular mixing pattern of the jet in the same way that mechanical suppressors had modified the mixing process of the early jet engines. However, although the spades were effective at model scale and in static engine tests, they did not reduce jet noise sufficiently under flight conditions to justify the thrust loss that they incurred.

The concept of spades in the exhaust system (to accompany the buckets!) had been one of the outcomes of the extensive programme of work undertaken by the four companies, supported by the advice of the group of academic experts drawn together in the Jet Noise Panel. Considerable theoretical and experimental effort was expended in trying to understand the mixing process of the high-speed jet and identify the major noise sources, and various concepts for promoting rapid mixing were evaluated. Unfortunately, most were either just too inefficient aerodynamically or suffered a large reduction in their effectiveness as noise suppressors in the flight, or "forward speed" régime,[26,836] a problem that was never laid to rest during the life of the development programme.

Although many devices were explored, there was never any question of a permanently deployed conventional jet suppressor replacing the complicated variable nozzle system that was needed to control exhaust flow under the wide range of operational conditions the aircraft had to satisfy. The nearest approach to a conventional suppressor nozzle was a device that allowed a small flow of nacelle cooling and ventilating air to bleed into a series of chutes upstream of the final nozzle. This was actually tested at full scale (Fig. 5.6) but found to be ineffective, whereas another proposal, to utilise a variable-plug system (Fig. 5.7), only reached the model-scale test stage. In this case, noise suppression was to be achieved by temporarily corrugating the final nozzle into something approaching the geometry of a conventional suppressor nozzle by using flaps that rose from the "plug" centre body. Nozzle area would have been varied by a moving shroud and a series of trimmers positioned between each flap, which were capable of rising and falling to modify the size of the plug. From the standpoint of noise suppression, the nozzle was as effective as any device tested at model scale but, from the engineering standpoint, it was deemed to be too complicated and likely to present unacceptable weight and cruise-drag penalties. The achievement of full pay-load and range were of paramount importance to the viability of the project.

It was for these reasons that the experimental work concentrated on devices that could be retracted and stowed out of harm's way for all but the take-off and early climb phases of operation. In this way, performance losses during almost all of the flight envelope would be avoided, except for the small weight penalty of the additional metal being carried. Most of the devices tested took the form either of simple flow disturbances ("fingers") or lifting surfaces ("spades") that would

Figure 5.6. Typical full-scale experimental noise suppressor for the Concorde.

disrupt the normal ordered pattern of the mixing process downstream of the final nozzle. Many variations on the theme were explored at small scale, usually under cold flow conditions, and the noise tests were often supported by flow visualisation techniques, as an aid to understanding the total process. One visualisation technique involved was Schlieren photography, examples of which are given in Figure 5.8. This shows how the shock cells in the supersonic flow from a round nozzle are disturbed by the insertion of a single "finger" downstream of the nozzle.

This type of disturbance was known to be a powerful means of eliminating shock-associated noise, but it also needed to have an effect on the generation of mixing noise. Hence, a whole array of nozzle inserts was tested over a period of about twelve months but, unfortunately, most would not have materially affected the noise from the Concorde exhaust system without incurring unacceptable thrust loss. The spades that found their way into the engine development programme[26] were the most effective of the devices tested. Even so, the amount of noise reduction they achieved was a function of the degree

Figure 5.7. Proposed variable-plug suppressor nozzle for the Concorde.

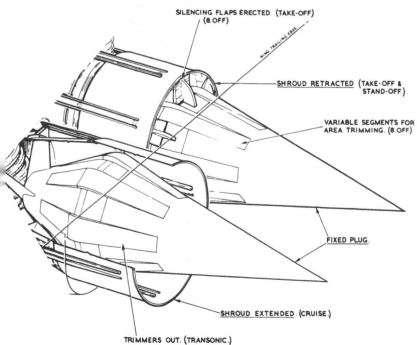

of deflection that they imparted to the flow, the parameter that also dictated the amount of thrust loss, and the spades only reduced the noise by some 1 or 2 dB before causing unacceptable thrust loss. Figure 5.9 shows how noise reduction varied with nozzle area and

Figure 5.8. Schlieren pictures of a jet with a finger inserted in the flow downstream of the nozzle (0.5-μsec spark on left, 0.01-sec time exposure on right): (a) undisturbed jet; (b) single finger, top view; (c) single finger, side view.

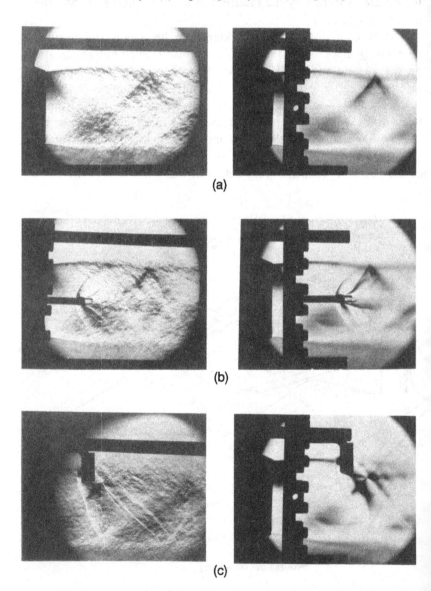

spade angle and how the thrust loss rapidly became 5–10% of that available – clearly, too much to throw away where performance is critical on take-off.

All in all, the Concorde programme revealed just how difficult it is to reduce the noise of a high-velocity hot jet. At the end of 1972, the four companies informed ICAO of the progress in reducing Concorde noise, for ICAO was still deliberating whether to include the supersonic transport in its evolving suite of certification requirements. ICAO was told that about 125 theoretical acousticians and practising engineers were studying the problem and that full-scale engine noise testing had topped 500 hours, or nearly 100 000 noise measurements. Flight testing on both the Concorde and the Olympus–Vulcan flying test-bed had covered 275 flight conditions, or 2000 noise measure-

Figure 5.9. Variation of peak perceived noise reduction and percentage thrust loss with primary nozzle area and spade angle.

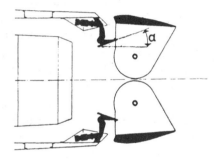

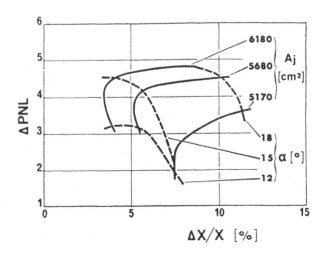

ments, and the total expenditure in facilities, testing and brain power had been some $100 million – a figure close to half of the original cost estimate for the whole project, but an effort that was considered "appropriate to the scale of the project". What a tragedy that the effort was all to no avail, other than the lessons it provided for the future.[47]

5.4   Post-Concorde

A revival of interest in the question of supersonic transport noise, after the Concorde had entered service, was sparked by ICAO's thoughts on future SST noise legislation and led to further experimental work. This included an extremely serious evaluation of the ejector concept, discussed in some detail in Chapter 4.

Note that a translating ejector was seen as probably the only realistic mechanism for reducing the jet noise from a high specific-thrust engine, if this cycle was to be selected for a future SST. Noise reductions of about 10–15 PNdB were predicted to be possible, with "acceptable" thrust losses on take-off. Acceptable, in this context, was about 1% for every 2 dB silencing. In the climb and cruise mode, with the ejector "passive" in its retracted position, the system was seen as not having to worsen fuel consumption by more than about 2% or, typically, far less than half the likely percentage take-off thrust loss.

An alternative approach was considered in the United States, which relied on a bypass engine cycle but had the hot jet on the outside of the cold flow. Known as the "inverted velocity profile" concept,[365] it was actively pursued as a research concept but did not achieve suppression values as great as the ejector and was dropped when the second steep rise in fuel prices of the 1970s pushed all thoughts of a new SST into the background.

When ICAO conducted its multination evaluation of possible future SSTs and their noise characteristics,[804] a less optimistic suppressor performance penalty was assumed (Fig. 5.10). Although the main conclusion was that the original subsonic jet noise certification standards of Annex 16 Chapter 2 "might be attainable", since then, and despite the good ejector results[837–9] and thoughts of "variable cycle" engines,[22] nobody has claimed that a new SST could achieve the latest Chapter 3 standards.

Not surprisingly, therefore, for this and economic reasons, there has been little real interest in civil SSTs since the 1970s. However, the recently declared U.S. "goal" of high-speed flight (the so-called Orient Express) and the European Horizontal Take-Off and Landing project

Figure 5.10. Jet suppressor performance – noise versus gross thrust loss (summary chart from ICAO Circular 157-AN/101).[N14]

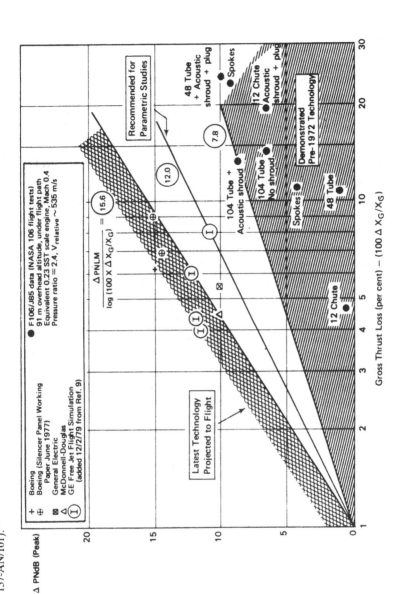

(HOTOL) are beginning to make industry and government rethink the possibilities.[51] Even so, it is widely acknowledged that any civil development of these will have no effect on the airport noise issue until well into the next century, for no "new technology" has been developed that would make the SST compatible with its subsonic counterparts. About the only thing that has changed is the group of people studying the options – a new generation of engineers, perhaps trying to reinvent the wheel. The fundamental problem of high-velocity jet noise[47] has not gone away. And it has to be solved, or avoided in radically new engine design concepts if the SST is to ever "get off the ground" and compete side-by-side with the quiet subsonic fleet.

To emphasise the point, Figure 5.11 provides a simple lesson from the past. This illustration expresses the design cycle of all the major engines that have been produced in the past twenty years as a function of "fully mixed" jet velocity or, simply, thrust divided by massflow. Aircraft with engines having a jet velocity of 500 m/s or more have either failed to achieve the original noise certification standards or, when close to the limit, have managed to scrape by with a hushkit. Where the jet velocity has been below 400 m/sec, the latest certification standards have been met, with due attention to nonjet noise sources. Who then would be bold enough to contemplate a new SST with engines having a jet velocity of over 400 m/sec at the take-off condition, or bypass ratio of less than 2, unless they were confident of

Figure 5.11. The message from the past – bypass ratios and fully mixed jet velocities of a range of commercial engines.

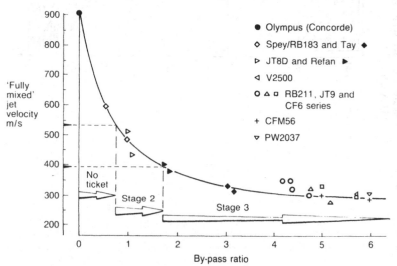

designing an effective jet suppressor that had aerodynamic performance vastly superior to anything seen before? Even then, it should be
remembered that the PW JT8D re-fan engine,[845] which falls on the
400-m/sec boundary, is being cited by some communities[29] as a bad
example of a Stage 3 engine and one reason why government should
be looking at even tougher noise legislation.

5.5   Summary
The Concorde was an exciting technical venture. However,
from the commercial standpoint, it was recognised as not being viable
by the time the development programme had reached the midway
point. However, the two collaborating nations judged that there was
sufficient technical and political justification for completing the project
and, to their credit, saw the programme through despite dire economic
problems at the time. It was no fault of the designers of the Concorde
that technology, in the form of the high-bypass engine, overtook them
or that market forces and world conditions drove the price of aviation
fuel to five times what it had been at the start of the programme.

From the noise-control standpoint, however, the Concorde was a
total failure in all but the exploitation of the variable-geometry features of its power-plants and low-noise operating techniques. No practical solution emerged to the extremely high levels of noise produced
by a reheated "pure" turbojet engine cycle. The hope for the future
must lie in variable-cycle technology, which is only in its infancy.
Despite claims, supported by limited demonstrator testing, that mechanical suppressors of the retractable ejector type could achieve noise
reductions of 10–15 PNdB, the aerodynamic performance has not
been evaluated for either the complex installation geometries associated with SST power-plants or the critical high-speed cruise condition.
Moreover, unless the variable-cycle engine concept proves viable, it is
quite possible that well in excess of 15-PNdB noise reduction will be
necessary for the next generation of supersonic transport if it is to be
compatible with the high-bypass ratio subsonic fleet that will be domi-
nating the aviation scene by the end of this century.

The Concorde taught us many lessons, a most important one being
that people cannot afford to travel hopefully when faced with a fundamental technical problem that is not understood at the start of the
programme.

6

Noise data acquisition and presentation

Aircraft noise and related data are acquired for a multitude of purposes but, broadly speaking, they fall into two categories – those for public presentation and those for research experiments. The method of measurement and the systems involved in processing data for public consumption are normally closely controlled, with a range of specifications determining the instrumentation system standards and the way in which data are acquired and analysed.[900-13] The most rigid form of measurement control occurs in noise certification.[1-7,50] Since the certifying authority does not itself conduct aircraft noise tests, it is the task of the aircraft manufacturer to comply with all the requirements and present the information in a standard form. The equipment and provisions made by the manufacturer are closely scrutinised by the authority for compliance before and during the tests.

The formulation of certification specifications has been a monumental and evolving task that has been going on for the best part of two decades and it is still unfinished owing to the continually changing state of the art. Many thousands of man-hours have been consumed in the task of refining international standards, with the important objective of reducing the variability of results obtained by different organisations that use different equipment and operate in different climatic conditions around the world. The necessity for such stringent specifications is now being amply justified as aircraft noise requirements at the national and local level creep ever closer to industry's capabilities.

In the mid-1970s, new wide-bodied aircraft fitted with the then-new high-bypass-ratio engines were clearing the noise certification requirements established in 1970 by as much as 5–10 EPNdB, so that procedural inaccuracies of the order of 1 EPNdB were of no great significance. With the progressive increase in stringency since that date, many aircraft manufacturers now seeking certification do so with an eye on the more demanding local airport rules,[10-12] where every tenth of a decibel counts. Confidence in the measured data is paramount.

To date, no aircraft has actually been prevented from entering service by virtue of failing its certification test, but that situation could occur eventually. However, as a result of failure to comply completely with the certification rules, several aircraft have been restricted in their day-to-day operations (in terms of maximum take-off weight or flap setting and hence field length on approach), all for the sake of the odd decibel or less. Operational restrictions at the local level are common. At one U.S. airport, the trade-off runs as high as twenty passengers (or around 2000 kg fuel or freight) for a change of one decibel! This is because the monitored limits are only influenced by increasing aircraft altitude or reducing engine power over the community – the effectiveness of both measures being a function of take-off weight.

It is worth pausing to consider the history of aircraft noise measurements. During the 1950s, both the systems and mode of acquiring and analysing data were comparatively crude, but there were no certification requirements and the accent was on demonstrating the change from one engine standard to another.

During the 1960s, more and more organisations were making measurements, with the result that valuable work on systems and atmospheric effects was conducted. Inevitably, differences in techniques and methods of analysis existed and, as the build-up to noise certification gained momentum, the need for standardisation became apparent. A programme of work was initiated to determine the magnitude of measurement differences, identify potential causes and make recommendations for improvements. A special test exercise[928] was conducted in California in which eight organisations took part and made simultaneous measurements of stationary and flying aircraft. The test served a useful purpose in that it demonstrated that results were strongly dependent on equipment and techniques. In 1968, a precertification "dummy run" was organised in Spain using a BAC 1-11 aircraft and, with more control of procedures, the standard deviation of over twenty results was shown to be less than 1.5 dB. By 1969, the FAA laid down its procedures in the draft of FAR Part 36 of the Federal Aviation Regulations and, in 1971, ICAO's Annex 16 to the Chicago Convention on Civil Aviation addressed the same issues. These early specifications benefited from the lessons learned in precertification work but further comparative work involving thirteen industry and government organisations exposed the shortcomings of the analysis process, particularly in the high frequencies, where ambient/ system background noise can have a marked effect.

By the mid-1970s ICAO had called upon the International Standards Organisation (ISO) for assistance and subsequently asked the International Electrotechnical Commission (IEC) to consider the whole question of procedures and practices for noise certification. This work laid the basis for the latest procedures of Annex 16 and, although no more intercomparative exercises have been launched to ascertain the value of the latest refinements, measurements conducted by aircraft and engine companies are now expected to be intercomparable to about 0.5 dB.

Modern systems are moving rapidly into the digital arena, with results from engines operating on open-air test-beds showing enormous improvements over preceding FM equipment. Standards organisations are already working on specifications for digital equipment, both for these purposes and for measuring flyover noise.[2] One can confidently predict that digital systems will be a way of life in determining aircraft noise by the mid-1990s.

6.1   Aircraft noise measurement

For definitive noise measurements, it is universal practice for the electrical signal from the microphone to be recorded on magnetic tape for subsequent "off-line" analysis. Although it is technically possible to analyse noise and compute EPNL in real time (and some monitoring systems do just this), in a noise certification test a number of adjustments have to be made to the data to correct them for electroacoustic-system response, for reference (standard) atmospheric conditions, for deviations from the reference flight-path and any variations in engine power setting. To perform these adjustments with accuracy it is necessary to have a detailed $\frac{1}{3}$-octave-spectral analysis throughout the noise–time history, and the most convenient way of doing this is to perform the analysis subsequent to the flight test.

The main items of equipment involved in an aircraft noise measurement exercise are as follows:

a microphone system, a calibrated amplifier and a tape recorder;

a means of relating the aircraft position to the noise recording; and apparatus for determining temperature and humidity conditions over the whole propagation path between the aircraft and the microphone and windspeed close to the ground.

The microphone employed is usually an air condenser type with good stability and a simple cylindrical shape. The so-called half-inch (diameter) microphone is normally used as it has a flat frequency response and is virtually omnidirectional over the audio range when it is used under grazing-incidence conditions, that is, with the plane of the diaphragm vertical as the aircraft passes directly overhead (see Fig. 6.1). This eliminates corrections for changing sound-wave incidence with respect to the microphone, which would be needed if the microphone diaphragm were horizontal.

Figure 6.1. The standard 1.2-m-high grazing-incidence microphone configuration for aircraft flyover noise measurement.

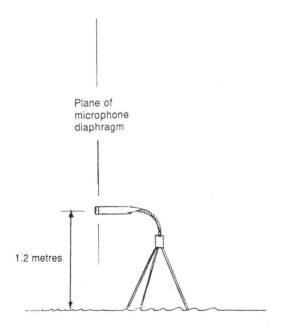

Plane of
microphone
diaphragm

1.2 metres

The specification for performance of the tape recorder can usually be met easily, provided it is a modern instrument using frequency modulation or even a high-quality audio machine, applied in the direct recording mode. The latter is inferior in linearity and calibration stability, but tends to be somewhat better in dynamic range and can "capture" sharply falling high-frequency characteristics.

Various methods of ascertaining aircraft position are employed, for example, photographically (Fig. 6.2), ground-based self-tracking laser systems and kine-theodolites, or even aircraft-mounted systems. Currently it is necessary to use an independent ground-based system for certification purposes but consideration is being given to accurate airborne systems. Atmospheric temperature and humidity measurements are made by means of aircraft-borne instrumentation (although these are usually on a separate aircraft) or by free-flying or tethered radiosondes (Fig. 6.3). A tethered system has the advantage of providing wind data although, as yet, such information is only used to indicate the presence of abnormal conditions such as severe wind shear or turbulence in the area of the test site. Noise, position tracking and atmospheric data recording are usually drawn together by using a common timebase transmitted by VHF radio, or landline, to the various recording centres.

Figure 6.2. A typical optical aircraft-tracking instrument.

Once the flight-test exercise is complete, the computation of effective perceived noise level (see Appendix 4) requires a knowledge of the $\frac{1}{3}$-octave-band levels over the whole time history, normally at time intervals of one-half second, with a running integration time of one and one-half seconds. A fundamental difficulty in the analysis and a primary cause of variability lies in the character of the noise signal, which is a nonstationary random signal and usually contains prominent Doppler-shifted tones. The determination of band levels at half-second intervals demands analysers having true RMS detectors and filters of known effective bandwidth. Modern analysers now meet this requirement satisfactorily, but for many years the detector characteristics were suspected to be a primary cause of variability.

After almost two decades' experience, aircraft noise measurements have reached a high degree of standardisation and reliability. However, two problems remain unresolved, when require attention if variability is to be further reduced. These are the location of the measuring microphone and the effect of atmospheric inhomogeneities in the propagation path. The question of microphone position is dealt with later in this chapter, but the problem of atmospheric variations is briefly discussed here.

To allow for any inhomogeneities in propagation, temperature and humidity values have to be obtained along the whole sound-path so

Figure 6.3. A weather-recording data pack suspended from a tethered balloon.

that adjustments in atmospheric attenuation can be calculated by "layering" the atmosphere into horizontal sections. However, the effects of wind shear, turbulence and, to a lesser extent, inversions have not been adequately identified nor have correction processes been defined. It is normal practice to try to avoid unstable conditions by monitoring aircraft speed changes and relying on the "seat-of-the-pants" feel of the pilot and the certification authority observer on board. This approach generally provides consistent data. Figures 6.4 and 6.5 illustrate the situation.

Two flyover recordings of a turbojet-powered aircraft were taken on successive days, with temperature, humidity and windspeed well within the limits specified for certification-test purposes. The pilot reported some slight turbulence. However, the tests were completed but the data exhibited a marked contrast in character from day to day. Both illustrations show a simple comparison of the overall sound pressure level (OASPL) against time together with a "three-dimensional" frequency versus time versus sound-level display, where sound level is given as the depth of shading. Immediately apparent in the 3-D plots is the "thumbprint" of the regular augmentation and cancellation pattern produced by ground-reflection interference effects. However, this is not the point of the illustrations. Rather, the important issue is the variation of the noise signal with time on the day when turbulence was present. Under the good, stable atmospheric conditions of Figure 6.4, the noise–time history is smooth, and the 3-D picture complete. With turbulence present (Fig. 6.5), the disturbed atmosphere causes "scintil-

Figure 6.4. Intensity–frequency–time recording of aircraft flyover noise in a homogeneous atmosphere. Density of shading shows amplitude of sound.

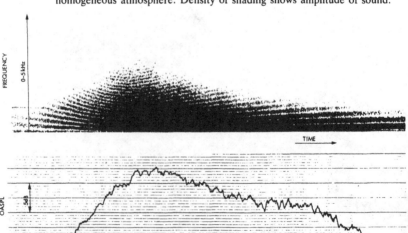

lation" in the signal, which affects the recorded level by over 10 dB at times. In theory, if the effect results from refraction, there should be both positive and negative changes around the true signal as the sound-wave is either scattered or focused but, in practice, the spreading effect of scattering dominates, probably due to upwards scattering, and even lengthens the total duration of the event. The true integrated energy level is more nearly equivalent to a line through the peak values on the OASPL plot, rather than a mean through the "fuzzy" time history.

Scintillation effects are not controllable – they are merely avoidable, provided the characteristics of the atmosphere are recognised ahead of the test. On very bad days, the effect is abundantly clear to an observer on the ground, since the noise appears to come and go in a random fashion as the aircraft flies by. It is the moderate case that causes problems, for it may only be detected when analysis is started well after the test has taken place.

On the other hand, the ground-reflection effects, which are so evident in Figure 6.5, are avoidable. A number of steps have had to be taken to provide meaningful research data as opposed to data for public consumption. In short, this means avoiding ground-reflection effects at the testing stage by using higher-quality data acquisition methods.

Research testing on model-scale component rigs avoids the problem because it is normally conducted in an anechoic environment free from reflected signals. This is not possible in the case of a flight test, for the ground surface is an ever-present reality. It would require the micro-

Figure 6.5. Intensity–frequency–time recording of aircraft flyover noise in a turbulent atmosphere – the "scintillation" effect. Density of shading shows amplitude of sound.

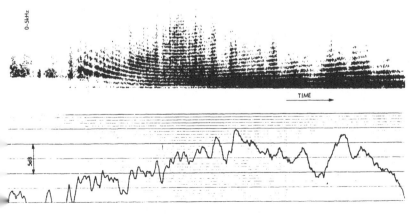

phone to be almost halfway between the ground and the source to eliminate, for all practical purposes, the ground-reflection interference effect. The nearest any organisation has come to this is in mounting microphones 10 m above the top of a 150-m-high bridge tower.[927] Normally, however, if there are no prominent low-frequency tones present, it is sufficient to position the microphone around 10 m above the ground, since all the major reflection effects are driven to frequencies below those of interest.[917] Unfortunately, this is not sufficient to solve the problem in the case of propeller-powered aircraft and helicopters, because they generate dominant low-frequency tones that are still significantly affected by ground effects.

The ideal way to eliminate the augmentations and cancellations resulting from the ground effect is to mount the microphone flush in an infinite and acoustically hard surface, so as to achieve full pressure-doubling at all frequencies; but, again, this is not a universally practical solution. Unless there is a hard runway or other similar surface available, the next best thing to do is introduce a "manageable" local artificial hard surface and to locate the microphone in the most practical manner to obtain near-pressure doubling over a frequency range of interest. Considerable research work[917-26] has led to the conclusion that the simple, "inverted", plate-mounted microphone configuration illustrated in Figure 6.6 represents the optimum balance between the technical and the practical considerations surrounding the conduct of a flight test involving small propeller-powered aircraft.[926] This conclusion resulted from field tests using steady-state artificial signals, a

Figure 6.6. A plate-mounted inverted microphone assembly as recommended by ICAO for propeller-powered light aircraft noise certification testing.

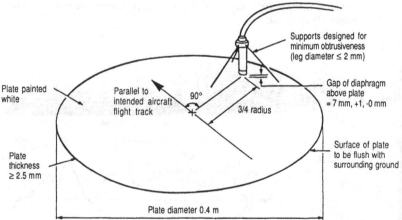

variety of aircraft and independent steady-state laboratory tests to evaluate the effect of varying the separation between the microphone diaphragm and the hard plate.[920] The laboratory tests showed that as the gap between the microphone and the plate is increased, reflection effects occur on a micro scale and begin to move down into the frequency range of interest in measuring aircraft noise. Conversely, if the gap is reduced too much, there is a "resonance" effect and an amplification in the higher frequencies, which is probably as big a problem as interference effects at large separations.

Both effects are illustrated in Figure 6.7, which summarises the response of an inverted system by reference to a flush microphone. This indicates why a diaphragm-to-plate separation of 7–8 mm has now been recommended by the SAE[926] and approved by ICAO's Committee on Environmental Protection as the international Certification Standard for the purposes of measuring the peak dBA noise levels of propeller-driven light aircraft. However, until it has been established in the field that this system is appropriate over the full range of frequencies and sound-wave incidence angles necessary to compute EPNL, the 1.2-m standard is likely to remain in force for all other types of aircraft.

6.2 Static rig and engine test measurements

Whilst the specifications and systems used to produce better repeatability and confidence in aircraft flight noise measurements are noteworthy, the situation with respect to measurements of engines undergoing static tests on open-air facilities is even better. It is because of the remarkably consistent results now being obtained from modern test facilities that static test measurements are becoming a regular cost-saving feature of the noise certification process.[2] These consistent, steady-state, static measurements provide a far better quantification of small changes in noise levels (brought about by comparatively minor power-plant modifications) than flight tests can ever achieve. In being confined to a small volume of "airspace", static tests can be conducted under more rigidly controlled environmental conditions and, by virtue of the data being extensively time-averaged, the short-time perturbations that would be a major concern in flight tests are virtually eliminated. Figure 6.8 reproduces a typical example of the repeatability from separate tests conducted on different days, using the same engine notionally set to the same power conditions. The repeatability is better than 1 dB over almost the whole spectral range. Compare this with the problems of the flight test data in Figure 6.5.

Figure 6.7. The effect of microphone diaphragm/plate separation with an inverted configuration.

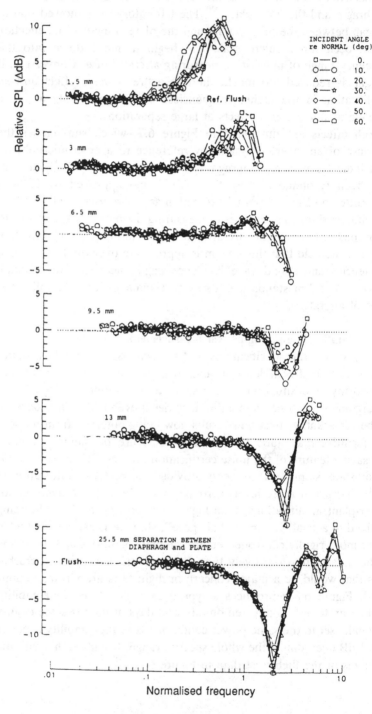

Figure 6.8. Typical spectra from repeated tests on the same test-bed but under different test-day conditions: (a) peak level forward of engine; (b) peak level to rear of engine.

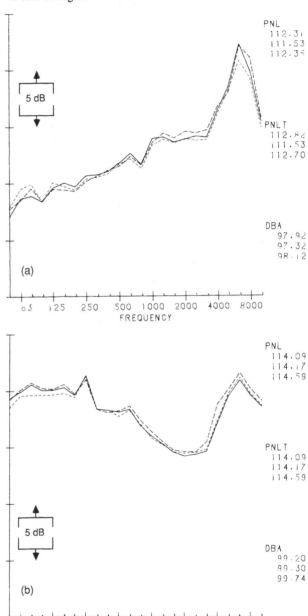

The static data in Figure 6.8 were acquired on the test facility illustrated in the photograph of Figure 6.9 (and earlier in Fig. 3.28), which is the latest in a line of purpose-built engine-noise test facilities, and which was specified by the author. It features a single-pillar support structure (to minimise reflection effects) mounted on a turntable (to permit the engine to be set facing into the wind on every test occasion). It also has a large inflow turbulence-control structure (designed to destroy the effect of large-scale atmospheric turbulence and more nearly simulate the flight condition)[914] and a multichannel microphone system feeding a digital record/analysis system that can handle up to sixty channels of data at any one time. Ground-reflection interference effects are eliminated by utilising inverted microphones over the extensive acoustically hard artificial surface, which is painted white to minimise solar-heating effects within the test arena.

Compare this operation to the full-scale engine test of twenty-five years ago. Figure 6.10 shows the microphone "array" for a near-field noise survey. The measurements are to be made over untended pasture-land, the accuracy of positioning being dictated by the engineer's ability to site one microphone over a wooden post whilst being bombarded by noise levels that resonate the skull and chest cavities,

Figure 6.9. Photograph of Rolls-Royce open-air noise test-bed.[921]

rapidly causing nausea. Those tests would have required half a day to achieve a survey of positions near the nozzle at, perhaps, 3 or 4 engine-power conditions. A week later, the $\frac{1}{3}$-octave analysis would have materialised, hand replayed and hand tabulated against visual observations of the replay levels. Today, the systems laid out on the hard arena of Figure 6.9 will allow control-room on-line analysis and display of several data channels and a comparison with data from the previous relevant test. Overnight, automated processing and correction of all the test data is conducted in both narrowband and $\frac{1}{3}$-octave formats. The "test- data" in this instance will be from twenty to twenty-five microphones every 5° around the engine at about 8 power conditions, acquired within an hour of testing.

6.3   Source location

Facilities like this, and those utilised for component research testing, rely on complex measuring modes and analysis techniques. These are necessary to improve our understanding of the complex noise-generating mechanisms. One such special technique is that of direct source location. Over the years, many attempts have been made to accurately define the position in space of the source of noises associated with the jet mixing process[931-9] and also to identify the main turbomachinery components.[940-3] In the far-field, these have ranged from directional microphones, to barriers and reflectors and the most successful technique – that of processing the phase relationship between signals recorded simultaneously at different points in space.

Figure 6.10. Engine noise measurement a quarter of a century ago.

Typical microphone arrays, often providing around thirty channels of information, are shown in Figures 6.11 and 6.12. With this technique, measurement accuracy is essential, for all data are recorded on a 2.5-cm-wide tape with the tracks only 0.4-mm width at around 0.5-mm spacing.

Figure 6.11. Microphone positioning for source location measurements on an engine.

Nevertheless, the technique has now been developed to an almost fully automated state, such that a computer-managed interpretation of the source description at a selected frequency can indicate its component entities. Figure 6.13 illustrates a typical result from a full-scale

Figure 6.12. Microphone positioning for source location measurements on a jet noise rig. (Crown Copyright – RAE)

Figure 6.13. Source location.

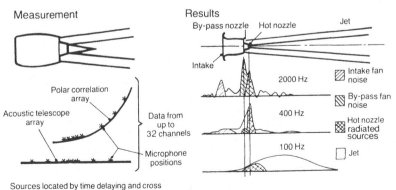

Measurement

Acoustic telescope array

Polar correlation array

Data from up to 32 channels

Microphone positions

Sources located by time delaying and cross correlating signals from an array of microphones to scan axial position

Results

By-pass nozzle   Hot nozzle   Jet

Intake

2000 Hz

400 Hz

100 Hz

Intake fan noise

By-pass fan noise

Hot nozzle radiated sources

Jet

engine test, with information being provided at three selected frequencies. At the lower frequency end of the spectrum, the sources within the jet are seen to be distributed downstream of the nozzle, whereas increasing frequency progressively reveals the importance of the internally generated sources – those from the fan, the turbine and combustor systems. Chapter 3.7 provides another example from a full-scale engine test, in Figure 3.42.

Not surprisingly, source location arrays, such as that shown in Figure 6.12, are a regular feature of some component research facilities. An example of their use is also given in Section 3.7, where a model of a modern high-bypass exhaust system was tested in a secondary flow system to simulate the effects of flight. As discussed, two distinctly different sources of jet noise were revealed, one close to the nozzle and one a significant distance downstream. Since this model did not contain any turbomachinery or combustion system upstream of the nozzle exit, source location clearly supported the theories propounding the presence of more than one significant noise-producing region in this type of jet. This kind of experimental technique adds credence to theoretical analyses, promotes a greater understanding of noise and assists in the development of predictive techniques.

6.4   Test facilities

Tests of the kind just referred to require more than good measurement and analysis systems; they require good test facilities. Over the years, a range of facilities has been developed around the world that will allow detailed examination of all the major component noise sources in the absence of extraneous influences.

In the 1950s, jet noise was a primary concern of manufacturers, government research establishments and universities in manufacturing nations, all of which boasted at least one, and often more, jet noise research rigs. These varied from (cold) compressed air systems exhausting into an open space through more elaborate indoor facilities (with temperature variability) to full-scale open-air engine test stands. As the science developed, there was a realisation that secondary noise sources, coupled with poor aerodynamics within many of these test rigs, were combining to contaminate the acquired data and that erroneous conclusions were being drawn. This was particularly true of the effects of flight speed on the mixing, shock-associated noise and other internal sources.

Comparison of data acquired by several major organisations, in an international attempt to agree on a harmonised jet-noise prediction procedure,[947] spurred efforts to improve test standards, to the point where there are now excellent facilities available, particularly in Europe and the United States. One such facility, at the Pyestock site of the Royal Aircraft Establishment (RAE) in England, is illustrated schematically in Figure 6.14, with the test chamber being shown in the photographs of Figures 6.15 and 6.16. This facility not only has the capability of testing single- and dual-stream jets, hot or cold, with simulated flight conditions via a tertiary flow, but it can also test small turbines, hot or cold. The important feature of this, and other similar facilities, is the attention paid to removing secondary sources of noise upstream of the test section. This is most important if reliable high-frequency data are to be obtained.

One of the issues that vexed jet-noise research workers for a long time was the apparently anomalous source effects resulting from changes in environment between static and flight test conditions, and with changes in flight speed of the aircraft (primarily up to around 0.3 M_n for community noise purposes, and even up to 0.8 M_n for cabin effects). Until researchers recognised that sources other than those

Figure 6.14. Diagram of RAE–Pyestock National Test Facility. (Crown Copyright – RAE)

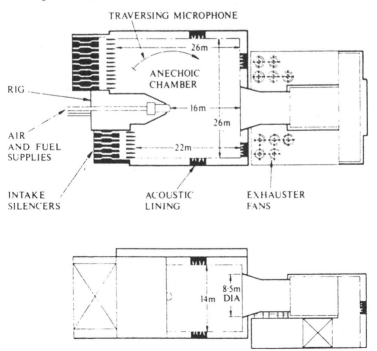

in the jet were having a major influence on conclusions, a range of "forward-speed" facilities was utilised (incidentally, what fixed-wing aircraft flies with "backwards-speed"?). Some manufacturers relied on aircraft flight tests supported by static engine data, whereas others developed unique facilities. In France, an experimental railway train[340-1] formed the basis of the speed corrections that found their way into one definitive prediction procedure; in the United Kingdom, a tip-jet helicopter rig[342] was used (see Fig. 6.17), with the data being averaged over several revolutions. In other cases, microphones were mounted on or trailed behind chase aircraft to acquire information on source variation with flight speed and on radiation directivity, in flight. It was not until work on aerodynamically "clean" wind tunnels[343-50] began to unscramble the various secondary source effects, and propagation effects in the moving wind-tunnel medium and its shear layer (where measurements were taken outside the tunnel working section), that the various other facilities fell into disuse. Nowadays, speed effects are usually examined in high-quality acoustic test sections of wind tunnels. These exist, for example, at selected manufacturers

Figure 6.15. General view of RAE–Pyestock National Test Facility anechoic chamber. (Crown Copyright – RAE)

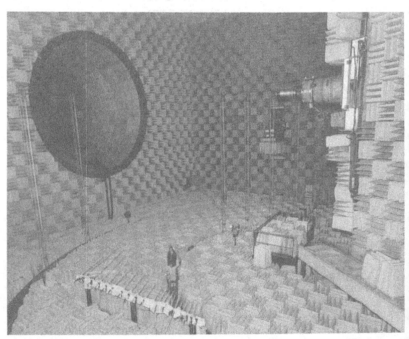

Figure 6.16. RAE–Pystock National Test Facility – a view from the exhaust duct. (Crown Copyright – RAE)

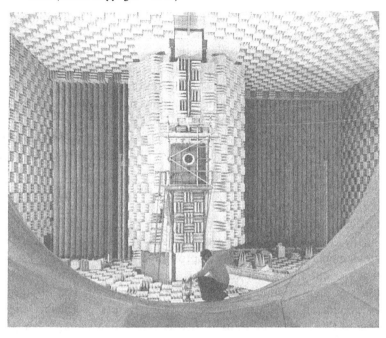

Figure 6.17. Photograph of Rolls-Royce jet noise "Spinning Rig".

(notably Boeing with its transonic wind tunnel) and government and industry establishments (e.g., the Dutch–German wind tunnel in the Netherlands, and the Aircraft Research Association and RAE in England). Although most of these facilities are being devoted to work on advanced rotors at the present time, established jet noise facilities have been equipped with additional flow streams to effectively echo the wind-tunnel environment at smaller scales.

The wind tunnels already mentioned, and others, have provided valuable data on conventional propeller and high-cruise-speed "open-rotor" performance in recent years, and they will be an essential feature of any developments in this arena. Open rotors, embracing both propeller and helicopter rotors, have to be tested in a moving medium, whereas the major components of ducted-rotor gas turbine engines, even full-scale power-plants, can be tested statically and adjustments made to closely approximate flight conditions. Provided that the entry and exit conditions can be made compatible with the relevant flight conditions, tests on fans, compressors and turbines can be conducted on static facilities. Manufacturers in Europe and the United States have all utilised such facilities, one example being that already discussed in the context of noise, the U.K. test facility at the RAE.

Perhaps the first major compressor facility to be commissioned was that owned by Rolls-Royce in the United Kingdom, which is illustrated schematically in Figure 6.18 and photographically in Figure 6.19. Driven by a steam-powered turbine unit from a salvaged destroyer sunk in World War II (which was the cheapest power source available when it was commissioned in 1966), this facility allows compressors up to almost a metre in diameter to be tested either exhausting into or obtaining inflow air from the anechoic chamber. In either mode, inflow air-conditioning is necessary to ensure that data are not affected by turbulence–rotor interactions (see Chapter 3). Interestingly, it was on this facility that much of the early work on inflow distortion was conducted, and it was unfortunate that, in common with many other facilities, several years' work on interaction tones was confused by inflow distortion effects.

One useful feature of the facility is that it has the capability to measure noise in three dimensions. The traverse boom, which can move between 0° and 120° to the inlet axis, carries microphones that can also move in a radial direction. A separate set of measuring equipment operates in the "azimuthal" mode, to provide full spatial definition.

Figure 6.18. Diagram of Rolls-Royce fan noise test facility.

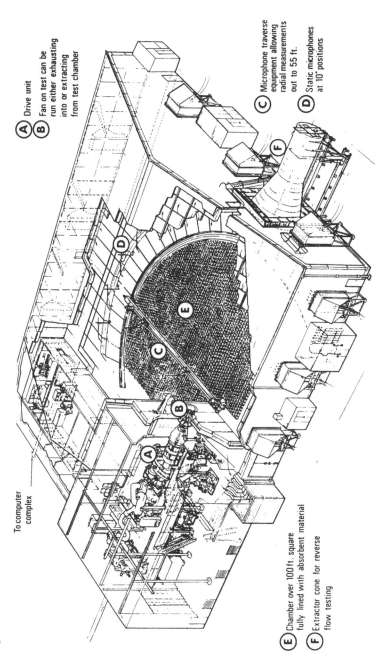

To computer complex

(A) Drive unit

(B) Fan on test can be run either exhausting into or extracting from test chamber

(C) Microphone traverse equipment allowing radial measurements out to 55 ft.

(D) Static microphones at 10' positions

(E) Chamber over 100 ft. square fully lined with absorbent material

(F) Extractor cone for reverse flow testing

Figure 6.19. Photograph of interior of Rolls-Royce fan noise test facility.

Figure 6.20. Photograph of RAE–Pyestock absorber facility test section. (Crown Copyright – RAE)

Other acoustic facilities used in the aircraft business include those for examining the performance of acoustic absorbers, for studying other sources (such as combustion noise) and for examining acoustic fatigue properties of materials that are going to have to withstand high-intensity sound fields.

Facilities for investigating performance of acoustic materials range from laboratory-scale impedance tubes to almost full-scale test sections, with flow, that can simulate both the cylindrical and annular types of engine-intake and fan-duct systems. Figure 6.20 shows the test section of the RAE absorber facility at Pyestock.

Work on combustion noise always tends to be confused by the presence of jet mixing noise in the exhaust system from the rig, where the noise measurements are normally taken. The use of cooled microphone probes within the combustion chamber is one alternative approach to data acquisition, but thus far it has not proved successful in determining either the level of or the leading parameters associated with the combustion noise process. Acoustic fatigue work is conducted in reverberant chambers with high-intensity drivers capable of simulating both discrete and broadband sources. With the advent of the much quieter turbofan engine, structural fatigue has become less of problem in parts of the civil aeroplane close to the engine, although it is still a concern in military aircraft having high pressure-ratio jets. However, methods of predicting fatigue characteristics have improved,[946] and the amount of experimental work has diminished accordingly.

For the reader who is interested in test facilities for aeroacoustic work, there is no single reference that will identify the wide range that exists. However, government departments in most noise-conscious countries are aware of their own, their industries' and overseas facilities.

6.5   Packaging for public consumption

Thus far we have briefly discussed some of the technical aspects of measuring and analysing aircraft-related sounds. Most of the technical issues are straightforward and any problems are usually resolved in the scientific community. Nevertheless, over the past thirty years, the apparently simple process of firstly measuring aircraft noise and then presenting the results to the public has posed significant and often unresolved problems in the pseudopolitical arena. These have ranged from the basic logistics of having to measure noise at diverse points around the aircraft flight track and relating these data to the aircraft's position in space and the pertaining atmospheric conditions, to prescribing the method of correcting data to standardised reference conditions so as to allow a "sanitised" set of numbers to be released for public consumption.

All the self-inflicted wounds arising from the cultivation of a thorny noise certification process have demanded treatment in the form of considerable investment in equipment and techniques, so as to improve the accuracy and confidence in measurement. Nowadays, the accuracy of the measuring equipment is expected to be good enough to produce repeatable results in terms of tenths of decibels, whilst overall EPNL repeatability in a flight test series can be less than a decibel. This includes all the variations due to changing climatic conditions, aircraft variables (including the pilot) and system inaccuracies. Static test repeatability is even better.

This state of the art has been achieved without significant change in the basic modus operandi. As discussed, recordings are usually made "in the field", on magnetic tape, subsequently analysed in the laboratory and then processed to allow data normalisation to reference conditions. The systems have become more accurate and reliable, with better signal-to-noise characteristics, and are now being improved further in moving from the analogue to the digital mode. Equipment manufacturers have developed a range of excellent systems to measure noise, some specifically aimed at the aircraft business. If any criticism is to be levelled at the overall method of aircraft noise quantification, it should be directed not at those responsible for the development of the electronic systems, but at those who have specified the way in which the systems should be used and then how data are parcelled and presented to the public.

It is here that we experience at first hand the impact of bureaucratic intransigence. In developing standards, multinational committees often agree on a "lowest-common-denominator" political solution to a technical issue and then subsequently fail to rectify basic errors when the opportunities arise because their actions must be palatable to the majority of the participants. Moreover, because such activity has its origins in public pressure, any decisions must also be "salable" to the public – that is, always assuming that "the public" are interested! As a result, several basic problems have been overlooked for some time. In brief, these problems are related to the way in which measurements are taken, the way in which data are corrected and the "reference" conditions to which they are corrected. Let us consider some of the most important ones.

6.5.1   Microphone positioning

The overall process of measuring aircraft noise to any defined standard is expensive and time-consuming. Hence, it is surprising to find that, after almost one-third of a century of aircraft noise measurement, the international standards still allow the microphone to be positioned in a manner that leads to the greatest possible contamination and confusion in the data so acquired. In aircraft noise certification, one of the most closely controlled noise-measurement processes, the applicant undertaking compliance-demonstration testing, in pursuit of a noise certificate, is instructed to place the microphone:

> . . . approximately 1.2 metres (4 ft) above . . . relatively flat terrain having no excessive sound absorption characteristics such as might be caused by thick, matted, or tall grass, shrubs or wooded areas.

ICAO Annex 16,[1] from which this quotation is taken, reads more like a gardening guide than a scientific document!

The reason for the 1.2-m specification is lost in the annals of time. On the one hand, it is rumoured to have been a misprint for 1.5 m, which approximates the height of the ears when one is in a standing position; on the other hand it may be an approximation of the ear height when one is relaxing, in a seated posture. Either way, it is nonsense, for few people actually listen to aircraft noise out in the open, seated on a park bench (well away from matted grass), with their biaural systems above an (almost) acoustically hard surface. Most aircraft noise is experienced inside a building, often with doors and windows closed, when sleep, conversation, thought or entertainment are disrupted by the sound signal that manages to transmit itself through the building structure. This situation has little or no relevance to the way in which data on aircraft noise are acquired.

The choice of a 1.2-m-high microphone position imposes a ground-reflection interference pattern on the actual noise data that is virtually impossible to unscramble in order to allow the results to be put to good scientific use. As illustrated in Figure 6.21, the direct and reflected waves from the aircraft noise source will interfere between the limits of pressure doubling and cancellation. The precise details of the interference pattern will be a function of the speed of sound (and hence day temperature), frequency, the source and receiver heights and propagation angles of the direct and reflected sound-wave. With the microphone at a height of 1.2 m, there are major cancellations in the lower $\frac{1}{3}$-octaves of the audible range, and the higher frequencies are affected according to the "softness", or acoustic impedance, of the ground surface. Most natural surfaces are absorptive at high frequencies, those with a dry fibrous texture close to 100%. Thus, the "excessive" clause of Annex 16 turns out to be nonsense.

Although it might be possible to correct for all these effects during the relevant period of an aircraft noise–time history if the ground surface were truly acoustically hard (i.e., totally reflecting), it is only possible to approximate the effect when measurements are taken over natural terrain of an unknown impedance.

Further complications arise as a result of a normal Doppler frequency shift in the aircraft noise source. All these effects are best illustrated by reference to Figure 6.22, where an aircraft flyover time history has been analysed in the same way as in Figures 6.4 and 6.5. Here, depth of shading represents the strength of the sound signal, the vertical axis indicates frequency and the horizontal base time. The ground-reflection interference pattern appears as the dark and light shaded pattern, with more sharply defined tones from the aircraft engine progressively decreasing in frequency (due to Doppler shifting) as the aircraft approaches and then recedes from the overhead position.

Since the interference pattern reduces in frequency as the aircraft is approaching, it is possible for Doppler-affected discrete tones to run either down a region of ground-reflection cancellation or a region of

Figure 6.21. Effect of microphone location on measured spectrum.

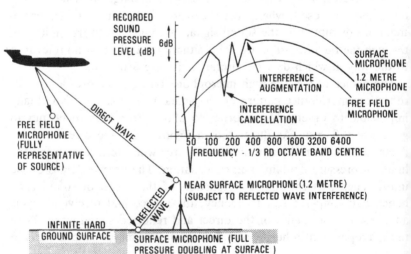

augmentation. The problem is less acute with the high-frequency tones of a typical fan or turbojet engine, but strong low-frequency propeller tones can fall either on a wave or in a trough of the interference pattern and either be augmented by 3 dB or completely "lost" by cancellation. Moreover, because of the narrow bandwidth of the lower $\frac{1}{3}$-octaves, one augmentation or cancellation can consume a complete band. As the aircraft flies away, the tones ride through the peaks and troughs in the interference pattern and the problem is less severe.

Figure 6.23 is a good example of the complete "loss" of the fundamental tone in a propeller-powered aircraft flight test, due to the interference cancellation of the 80-Hz band at a height of 1.2 m. Conversely, the second harmonic at 160 Hz is augmented, to almost the same level as that recorded with the ground-based systems.

Clearly, if data from an aircraft noise test are to be used for research purposes, different measurement procedures have to be invoked that avoid or, at least, minimise the effects of ground reflections. Over the years these have ranged, typically, from embedding microphones flush with an available (or artificially created) hard surface, to measuring at a height many times greater than the 1.2 m specified internationally. In fact, as discussed earlier in this chapter, some organisations have conducted experimental work at the top of bridge towers, others have used microphones at the top of 10-m masts and many have used microphones mounted in plates or boards, large and small.[917-27] Nevertheless, all data so acquired are inadmissible in computing the

Figure 6.22. The presence of genuine tones from a turbofan engine in the ground-reflection interference pattern from a 1.2-m-high microphone.

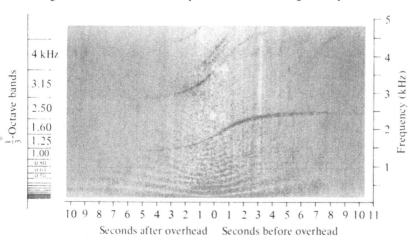

Seconds after overhead   Seconds before overhead

certification EPNL of the aircraft, for which the 1.2-m microphone height has been inviolate. Other data can only be "used in evidence" of the absence of a genuine tone and the consequent avoidance of a low-frequency "tone" correction due solely to the ground effect.

Whilst the "1.2 metres above . . . relatively flat terrain" requirement has caused many serious technical problems over the years, it has had its amusing side. One organisation, rehearsing their first-ever noise certification test, and operating at a local airfield in an area with several natural lakes, found the 6500-m take-off point was in the middle of a large stretch of water. Rather than going to the extra expense of moving their tests to an airfield much further away from home, they resorted to constructing a large raft. Even though the local humidity was fairly high and the raft covered with artificial turf to soften the surface on which an engineer was to have to squat for several hours, the whole contraption managed to conform with international requirements. After all, a lake is "relatively flat and free from thick, matted or tall grass, shrubs or wooded areas"!

On the test day, the raft was duly towed into the middle of the lake, and the most gullible engineer amongst the team was established "on

Figure 6.23. Typical turboprop-powered aircraft $\frac{1}{3}$-octave spectra from 1.2-m-high and ground-level microphones.

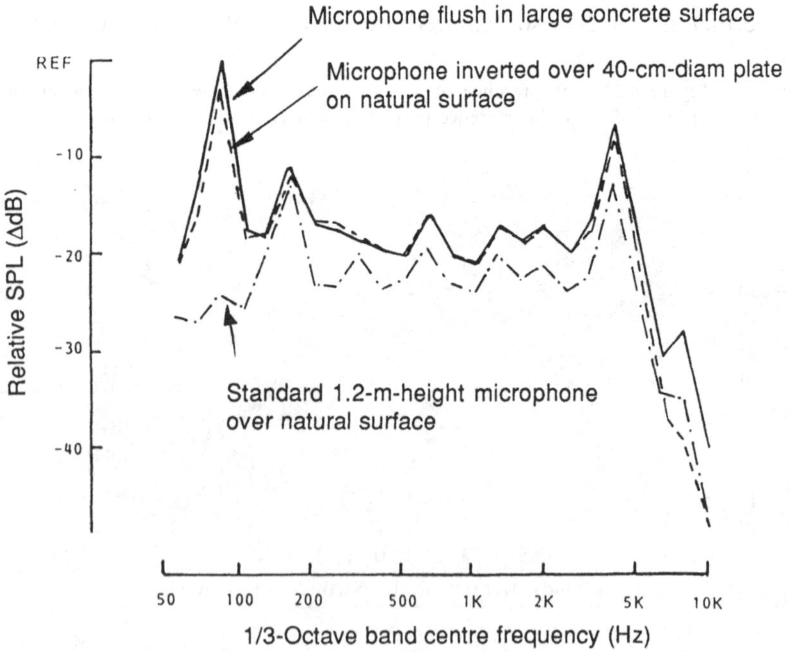

station". At the end of the test programme and in the gathering gloom, the sense of relief that all had gone to plan was so great that the main test team completely overlooked the recording engineer, who was very much alone and nearly invisible through the mist. Despite calls for attention, the communications system was closed down and the hapless soul on his unlit raft watched the team's hazy car headlights disappear. Fortunately, his absence was noted when the team sat down to dinner over an hour later – otherwise he might still be there!

Just what all this has to do with the 1.2-m-high microphone may well have escaped the reader. If so, it is worth considering just what relevance complying with certification requirements with a microphone 1.2 m above a carpeted raft in the middle of a lake has to the airport noise issue. The answer, of course, is nothing at all. So, perhaps, we should disassociate the mode of noise measurement for the purposes of certification from the everyday indoor–outdoor situation. This is what the scientific community has been pressing for over the past decade, and a chink of light is now beginning to show under the administrative door. As discussed earlier in this chapter, an alternative measurement system for propeller-powered light aircraft was recommended by ICAO's Committee on Aviation Environmental Protection (CAEP)[3] in mid-1986. This decision bore heavily on work conducted in the United Kingdom[922-5] during the late 1970s and early 1980s but, surprisingly, has been resisted by one major nation on the "NIH"* principle, even though they accepted the ICAO committee decision at the time.

The result of this political stubbornness is that manufacturers of light aircraft will have to do two certification tests, the extra one for the benefit of the dissenting state.

Worse still is the fact that the absolute noise levels reported from the two test procedures are likely to be substantially different, the difference varying according to the actual fundamental frequency of the propeller tone on the test day. As shown in Figure 6.24, for a propeller noise source with an 8-dB harmonic decay rate, as the tone frequency is varied (either as a result of the choice of blade numbers or changing rotational speed), there can be as much as a 7-EPNdB difference between the levels measured 1.2 m above ground, for a constant free-field source level. How, one wonders, will a difference of this magnitude mean anything to the "public", who are unlikely to understand the niceties of ground-reflection interference effects? Moreover, what would their reaction be if they realised that the aircraft designer could select a fundamental frequency that would give him the lowest cer-

---

* Not Invented Here.

tification value without affecting the intensity of the source, providing the test procedure was based on the 1.2-m microphone position!

6.5.2   *Correcting for atmospheric absorption of sound*

We spend our lives at the bottom of an ocean of air, our vital atmosphere. Sound is merely the progression of small pressure fluctuations in this atmosphere and, like the ground, the atmosphere absorbs sound. If this were not the case, we would still be hearing sounds that were generated many years ago. The rate at which sound is attenuated is a function of several factors, including two fundamental conditions of the atmosphere: its temperature and its water content. A third factor, source intensity, can confuse the issue at very high levels of sound, for the propagation can be nonlinear.[231] However, generally speaking, as both temperature (T) and relative humidity (RH) increase, sound absorption decreases. Absorption also decreases with a reduction in sound frequency, and below 1 kHz it is very small and almost independent of RH and T.

Over the years, theoretical evaluations combined with laboratory experiments have made it possible to quantify the effects of temperature and water content at any chosen frequency within the audible range.[222-30] In all cases, experimental studies have utilised a homogeneous atmosphere, and have dealt with the behaviour of discrete frequencies. Unfortunately, that section of the earth's atmosphere through which aircraft noise propagates is far from homogeneous, and its variability has a large effect upon the perception of the complex

Figure 6.24. Difference between true (free-field) propeller noise levels and those measured when using a 1.2-m-high microphone.

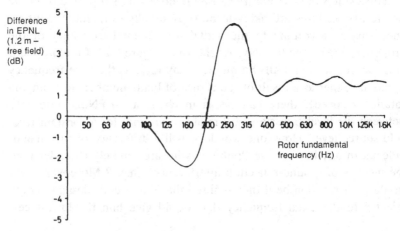

mixture of random (broadband) and discrete sounds that are generated by the wide range of commercial aircraft. From the technical standpoint, the situation is further complicated by the noise measurement and analysis process, which, for convenience, utilises $\frac{1}{3}$-octave (or broader) bandwidths. These bandwidths encompass ranges of frequencies that are subject to different degrees of atmospheric attenuation. Nowhere is the effect of atmospheric attenuation "flat" across a finite bandwidth.

Moreover, not only do temperature and humidity vary with changes in altitude (both decline as one gets farther away from the earth's surface), but the various effects that result from the rotation of the earth and the associated daily heating-and-cooling cycle produce conditions that are far from repeatable and extremely difficult to predict, even on a short-term basis. The noise heard on the ground as an aircraft departs from or approaches an airport has to propagate between the source and the receiver through the most unstable part of the atmosphere, that is, the region closest to the ground, where it washes the uneven surface of the earth and creates a turbulent boundary layer. The unstable nature of the medium can create large perceptible noise differences on a daily, and even on an hourly, basis. Equally, a given aircraft type can produce different long-term average noise levels at different airports around the world simply as a result of the wide variations in average local climate.

Agreed absorption characteristics of the atmosphere, for the purposes of correcting aircraft noise data, have been prescribed from the process of melding together the theoretical considerations with both laboratory and aircraft flight test data. These have been published in various forms,[222–30] for example, as in Figure 6.25. This set[229] of "agreed" characteristics has been relied upon for many years but, at the same time, has been criticised for not being technically accurate, and improved relationships[225] have been developed. Equally, the standard or reference condition to which noise data are corrected has also been criticised, from both technical and policy standpoints. Clearly, the vagaries of the world climate have to be accepted in the process of measuring, analysing, correcting and then presenting aircraft noise data in a harmonised fashion but, in so doing, everything possible should be done to ensure that only random errors influence the final answer. A statistically relevant statement of an aircraft's noise, or annoyance, has to be as appropriate to the residents of Bombay as it is to those of Paris or Los Angeles; otherwise the point of any harmonised statement or certification process is brought into question.

Therefore, although it is bad enough that decisions over the placement of the microphone create anomalies in the data recording, it is wholly unacceptable that allowance for the significant effects in the atmosphere is treated in an equally cavalier fashion. Let us look more closely at the situation.

Historically, the possibility of large variations in measured, and hence "reported", noise levels was recognised in the early days of aircraft noise measurement. At that time it was creditable to achieve about 3–5-dB repeatability in measurements of the same aircraft,

Figure 6.25. Typical atmospheric noise absorption chart (from SAE ARP 866).

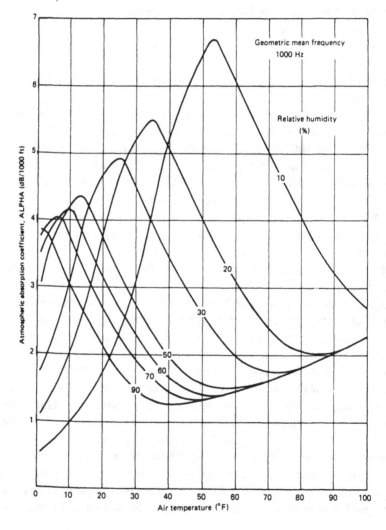

simply because of the inadequacy of the measuring systems and the lack of definition of harmonised test procedures. Correction for local atmospheric conditions was of secondary importance only, and it was not until noise-measurement systems and aircraft-position tracking techniques improved that the question of the atmospheric absorption of sound became a meaningful issue.

Once this point had been reached, an attempt was made to average out the many variables in the "real" (varying) atmosphere by allying the idealised laboratory data on single-frequency attenuation rates to the database of aircraft noise measurements acquired over a period of several years. This process made it possible to identify the main variables and incorporate the observed effects into a recognised set of standard atmospheric absorption tables.[229] These present the complex absorptive properties of the atmosphere on an average basis in convenient bandwidths ($\frac{1}{3}$-octave and whole octave). For this to be possible, however, the averaging process has to allow for the large variations in source spectra that characterise different aircraft types, the effects of changes in spectra with distance and, what is particularly important, the impact of conducting evaluations on a $\frac{1}{3}$-octave, rather than single-frequency basis. Widely different energy distributions exist even within a $\frac{1}{3}$-octave band, and these distributions change with increasing distance from the source as atmospheric absorption takes its toll on the higher frequencies. No single-number absorption value can exist that will reflect what happens in any higher $\frac{1}{3}$-octave band since the distribution of source energy varies from aircraft to aircraft, and then changes significantly according to distance from the source. Nevertheless, for "engineering" purposes some simplifying assumptions have to be made about the controlling band frequency (see Fig. 6.26). Generally, rapidly falling energy levels across the higher $\frac{1}{3}$ octaves are a feature of aircraft noise spectra and, here, the assumption is made that the lowest frequency of the band dictates absorption rates. Conversely, because source characteristics are generally "flatter" and atmospheric absorption effects are not so powerful in the low to mid frequencies, below 5000 Hz the centre frequency of the band is assumed to control. Unfortunately, these assumptions lead to a "kink" in the otherwise progressive change in absorption with frequency, which is of no importance when corrections are made to account for slight changes in day conditions at a given distance from the aircraft, but which can lead to large errors when measured data are extrapolated over large distances; for this reason, noise-contour computation takes a different approach to standardising attenuation rates.

Nevertheless, the assumptions outlined above are adequate for the purposes of data correction in a certification flight test, provided that the actual atmospheric conditions fall close to reference conditions and that the operation of the aircraft is closely controlled to provide minimum variation in flight tracks – in other words, provided that only small net corrections have to be made. It is when day conditions differ from those specified in the reference that problems arise. Then there is no hiding from the fact that the accepted reference atmosphere to which aircraft noise data are corrected is nonsense, and a good example of an early bad, albeit innocent, decision being perpetuated as an administrative convenience.

Those responsible for establishing the early noise certification criteria recognised that aircraft noise levels were reported differently in various parts of the world and, in particular, were higher when the

Figure 6.26. Selection of frequency to represent absorption characteristics of a $\frac{1}{3}$-octave band.

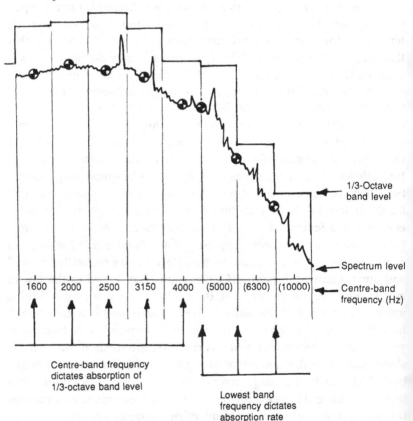

local climatic conditions were warm and humid. A DC8 operating out of Miami in July might well record a noise level over 5PNdB higher than the same aircraft operating out of Stockholm on a cold, dry February day. Any standardised certification test has to recognise this difference and provide for the correction of locally measured noise levels (and aircraft performance) to a harmonised reference condition. But, what should the standard reference condition be? The well-meaning thoughts of those wrestling with the problem in the early unenlightened days departed from good practice and, as a result, have left a legacy of overcorrected test data. Those thoughts were based upon the fact that certification was being developed to "protect the public", and therefore the "declared" noise values from individual aircraft tests should appear as high as possible. This was done so that the public could not claim that they were suffering higher levels around the airport than the levels published in certification data.

This meant that "minimum absorption" had to exist in the correction process. As a result, both a high temperature (25 °C) and relative humidity level (70%) were established as the reference conditions. Worse still was the fact that the atmosphere was prescribed to be homogeneous, when in reality both relative humidity and temperature vary considerably, on average, decreasing with increasing altitude (see Fig. 6.27). Hence, the majority of the reduction in sound that takes place naturally when an aircraft operates at an airport was deliberately excluded from the published certificated values, and artificially high noise values have regularly reflected the definitive production standard of a given aircraft type. The problem was actually recognised in the early days, for a special reference condition with a higher attenuation rate (15 °C, 60% RH) was incorporated into Annex 16 for use by countries conducting tests in a cold climate (e.g., the Soviet Union). However, the "declared" levels had to be adjusted upwards by 1dB when published alongside results from the rest of the "warmer" countries.

Even though the fundamental problem was registered early on, the correction process has been tampered with over the years in a way that has tended to make matters worse by further artificially increasing the published noise values. In demonstrating compliance with the original noise requirements, prevailing atmospheric conditions were measured by temperature sensors placed 10 m above the ground (Fig. 6.28(a)), and the conditions recorded at that position were assumed to exist along the entire sound-propagation path from the aircraft to the ground. In other words, homogeneity was assumed in both the test and

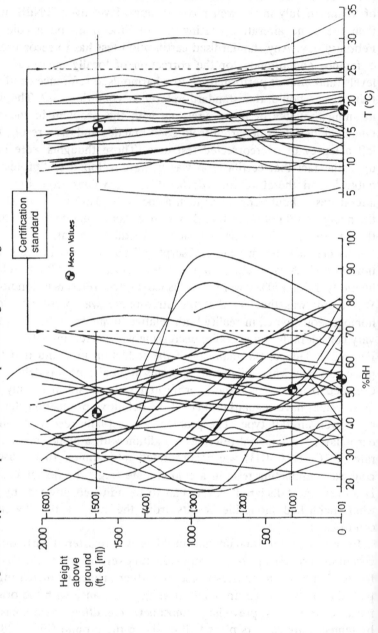

Figure 6.27. Typical atmospheric conditions reported during noise testing.

reference day cases and data correction was uniform over the entire noise propagation path. This allowed a simple "one-shot" application of the relevant correction. Subsequently, however, the correction process was supposedly made more "realistic" by a requirement to measure atmospheric conditions over the entire sound propagation path between the aircraft and the ground and to correct for all the varying differences in atmospheric absorption between the observed conditions and the artificial homogeneous reference atmosphere (Fig. 6.28(b)). Naturally, this move merely served to increase the "declared" noise value associated with a given aircraft type for, as Figure 6.27 has already illustrated, absorption rates increase as temperatures and rela-

Figure 6.28. The impact of the atmospheric layering process on correction of noise test data.

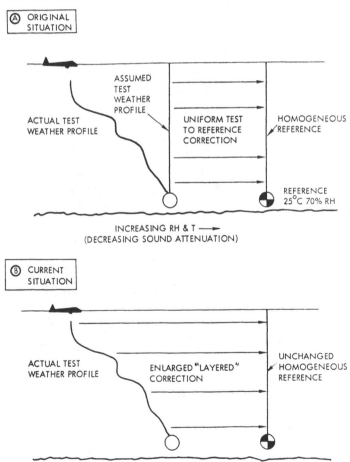

tive humidities decrease with increasing altitude. At the test altitude, the prevailing conditions can often be 10–15 °C and 20–40% RH lower than the homogeneous reference dictates they should be!

Figure 6.29 summarises the shortcomings well. It presents contours of constant sound attenuation rate (due to atmospheric absorption alone), within a matrix of temperature and relative humidity, for a frequency of 8 kHz. The reference condition of 25 °C, 70% RH

Figure 6.29. Measured test day conditions compared to Reference – the impact on attenuation rate at 8 kHz.

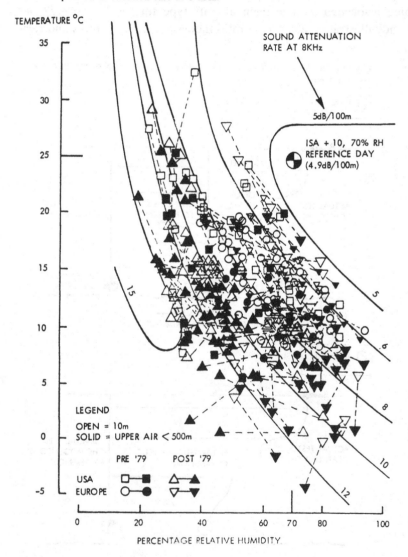

corresponds to an attenuation rate of only 4.8 dB per 100m, whereas actual atmospheric conditions measured during flight noise testing at a number of North American and European test airfields have rates two to three times higher. Actual atmospheric conditions never reflected the reference during the testing summarised, and in all cases the measured noise levels required upward adjustment, by a considerable amount; to conform with international standards: on average, some 3 dB per 100 m.

The true measure of disparity is better expressed in Figure 6.30, which offers a histogram of observed sound attenuation rates. This shows that noise produced during a take-off noise test conducted in

Figure 6.30. Histogram of 10-m and upper-air attenuation rates during aircraft noise testing: typical test day conditions reported from Europe and the United States.

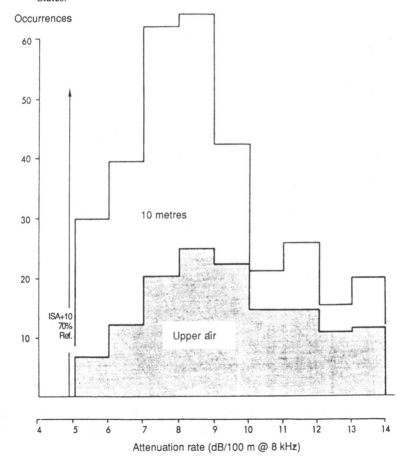

average conditions would experience an attenuation rate of about 8 dB per 100 m. This would mean that, for an altitude of, say, 600–700 m, the noise generated by a test aeroplane would suffer 8-kHz $\frac{1}{3}$-octave-band sound absorption of some 50 dB in propagation down to the microphone on the ground immediately beneath it. The reference condition would decree that a reduction of only 30 dB should take place, and the measured 8kHz band level should be adjusted upwards by some 20 dB. On cold, dry, test days the upward adjustments would be even greater.

Obviously, the correction of noise data from one condition to another places heavy reliance on the validity of the atmospheric absorption data, and the way in which it is applied to $\frac{1}{3}$-octave bands. Small errors are magnified many times over large distances and the accuracy of the atmospheric absorption tables and implied assumptions are of critical importance in determining the reliability and reality of noise levels reported from different test sites with different climatic conditions. Thus far, however, the only positive move to recognise the reality of average, rather than extreme, atmospheric conditions has come in the sphere of noise-contour predictions, for these have to relate to everyday monitored noise levels.

For many years individual nations had their own home-grown contour-prediction methodologies, but in the late 1970s it became increasingly obvious that this was only a recipe for one thing – confusion. Accordingly, there has been a coming together of minds on the subject, reflected in the harmonised noise-contour calculation process proposed by ICAO.[990] This method relies on a reference condition that is based not on an arbitrary temperature or relative humidity, but on the relationship between frequency and sound absorption as determined from the range of real atmospheric conditions at the world's airports. Figure 6.31 reproduces the relationship, as proposed but rejected by ICAO for certification purposes in 1983.

Clearly, it is one thing to think you know how wet or hot average conditions are, but quite another to appreciate the technical problems associated with decisions taken by amateur weather experts! It is to be hoped that, in the coming years, the wider use of the new reference absorption standard in noise-contour computations will persuade authorities around the world to adopt this standard for the purposes of certification and will lead them to examine the case for updating the standard atmospheric absorption tables in the light of modern studies of the physical relationships that dictate how sound decays in the atmosphere.

Figure 6.31. Reference day attenuation rates, as now used in contour computation.

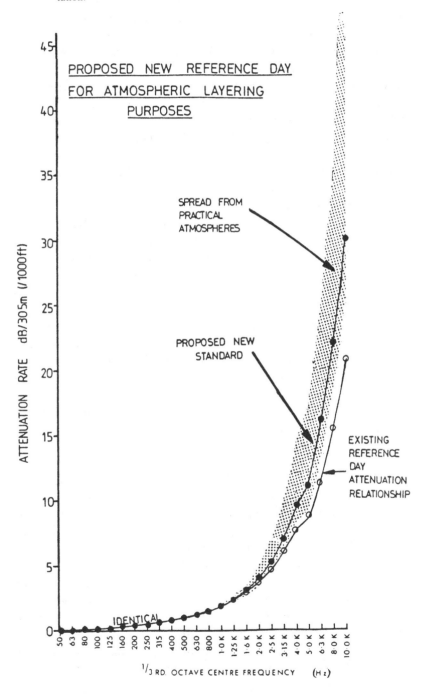

PROPOSED NEW REFERENCE DAY
FOR ATMOSPHERIC LAYERING
PURPOSES

SPREAD FROM
PRACTICAL
ATMOSPHERES

PROPOSED NEW
STANDARD

EXISTING
REFERENCE
DAY
ATTENUATION
RELATIONSHIP

IDENTICAL

ATTENUATION RATE dB/305m (/1000ft)

$^1/_3$ RD. OCTAVE CENTRE FREQUENCY (Hz)

6.5.3   The tone correction

The process of establishing a noise-annoyance rating scale has dragged for many years. As explained in Chapter 1, the basic noise level–frequency–annoyance relationship used in the calculation of PNdB had to be modified in the late 1950s following the realisation that college students had rather better hearing than the average citizen. Nevertheless, to this day, some exposure indices still use the original PNdB scale, the excuse being that the information that flows from government departments must be "consistent" from year to year! Hence, it can be appreciated that the development of the more complex effective perceived noise scale gave ample opportunity for "bureaucratic meddling". One such example is the approach to the topic of the tone correction.

A "tone correction" was first introduced into the process of annoyance description as a result of the annoying whine from engines like the earliest turbofan, the two-stage PW-JT3D. The tones from this machine often protruded well in excess of 10 dB above the $\frac{1}{3}$-octave broadband level, justifiably labelling the later Boeing 707 and DC8 aircraft as two of the most annoying aircraft ever produced. Laboratory experiments,[135–8] aimed at representing the presence of excessively protrusive discrete tones, indicated that these were more annoying than indicated by the basic perceived noise (PNL) scale. Accordingly, an additional annoyance factor or tone "penalty" was developed and applied to the basic PNL level, as part of the original effective perceived noise scale.[133]

The experimental evidence on which the tone correction was based was (and still is) somewhat limited, but at large tone protrusions it generally presented a consistent story. However, where the tone protrusion was small, it was extremely difficult to see any discrimination against tone-free spectra. Therefore, to avoid any argument, the early published work made the sensible technical judgment that no tone corection should be applied if the protrusion was less than 3 dB with respect to the local $\frac{1}{3}$-octave background spectrum level. This "threshold" was to apply to the most annoying frequency bands and a larger, 6-dB, threshold to less sensitive bands.

Although the 6-dB figure was never used, the 3-dB threshold, or "fence" as it has become known, was accepted for the purposes of the early noise certification scheme, and was used for computing EPNL. The relationship used is shown in Figure 6.32(a).

This correction stood unchallenged for many years, that is, until the tidy-mindedness of one national delegation led to the suggestion that

Figure 6.32. Effect of technopoliticking on scientific issues – the tone correction.

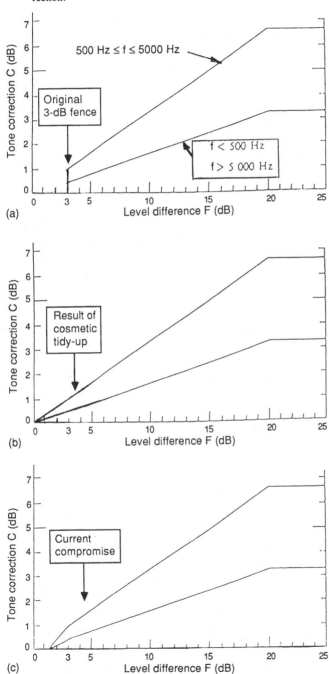

ICAO should delete the 3-dB fence and extend the tone correction down to the origin. The idea was seized upon by several members of ICAO's Committee on Aircraft Noise as a presentationally attractive, technically sensible, generally tidy and therefore a wholly acceptable "minor technical adjustment". The change was duly incorporated into Annex 16, as per Figure 6.32(b). The problem was that the matter had not been thoroughly considered before adoption, and it gave rise to tremendous complications the moment it was first used.

For example, the ground-interference peaks and troughs induced in using a 1.2-m-high microphone were, on analysis, construed by the system to be a series of discrete tones. These then required the application of a tone "penalty". Clearly this was utter nonsense. Previously, where there had been any low-frequency ground-reflection peaks and troughs that exceeded the threshold fence of 3 dB, they were generally ignored and discounted in the accounting process as "pseudotones". However, the deletion of the 3-dB fence immediately extended the range of pseudotones from the lowest $\frac{1}{3}$-octaves through the total frequency range, causing confusion with real tones in the mid to high frequencies and a whole new game of tone identification and accounting was set in motion. For example, the only genuine protrusive tone in the spectrum of Figure 6.33 is at 3000 Hz; all the discontinuities below 3000 Hz are the result of ground-interference effects at the 1.2-m-high microphone and should be discounted.

Even the standardised absorption properties of the atmosphere could lead to anomalous tone corrections! The reason for this was the discontinuity in the atmospheric tables, already discussed. When any smooth source spectrum is extrapolated using the standard atmospheric absorption tables, it develops a 4–5-kHz "kink". As Figure 6.34 shows, a smooth jet noise source was debited with a "tone" correction of around $\frac{3}{4}$ dB when it was extrapolated to a distance of 250 m using the standard absorption tables! This $\frac{3}{4}$ dB is due to the kink but, because of the second-order nature of the spectrum, a whole series of small tone penalties can also be computed.

Faced with this sort of anomaly, the rulemakers of ICAO were eventually forced to relent – but, predictably, not completely. It would have represented tremendous loss of face for some to revert to the original and well-tried 3-dB fence. So, there had to be a compromise, even though there had been no new scientific evidence on the subject. That compromise is shown in Figure 6.32(c) – a sloping fence, starting at 1.5 dB and mating in with the main slope at 3dB! Literally a "halfway house" solution, it is scientifically insupportable, but a good political compromise (that still produces erroneous penalties)!

6.5.4 The background noise problem – or, Consider the bumble-bee

For many years, the presence of background noise, whether it be genuine local ambient noise, wind noise on the microphone or electronic noise in the measurement and analysis systems, has presented problems in correcting measured data that are corrupted. If noise measurements were taken close enough to the aircraft there would be no background noise problem. However, they are not and, depending on the source character of the aircraft, its altitude and the properties of the atmosphere pertaining on the day of the test, local ambient and the electronic system noise can mask some of the signal. Genuine ambient noise (wind and other sources) affects low frequencies, whereas noise from the electronic system affects high frequencies (see Figure 6.35).

It is clear that some attention needs to he given to corrupted data, for the signal from a relatively quiet aircraft undergoing a noise cer-

Figure 6.33. Genuine and pseudotones in a high-bypass-powered aircraft spectrum.

tification take-off test can have up to ten $\frac{1}{3}$-octaves influenced by background noise, particularly near the "skirt" of the 10-dB-down time history. Over the years, several correction procedures have been proposed, all of which rely on an assumption about the nature of the aircraft noise signal before it was swamped by the ambient noise. As a

Figure 6.34. Example of broadband noise tone penalties.

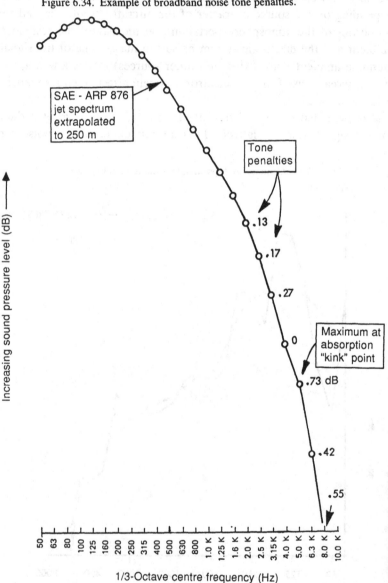

result, the correction processes are riddled with bureaucratic guide-lines. For example, Appendix 2, Section 3.5.6, of ICAO Annex 16 gives the prospective problem owner the following instructions:

> The ambient noise, including both acoustical background and electrical noise of the measurement system, shall be recorded at the measurement points with the system gain set at the levels used for the aeroplane noise measurements, at appropri-ate times during each test day. The recorded aeroplane noise data shall be accepted only if the ambient noise levels when analysed in the same way and quoted in PNL (see Section 4.1.3(a)) are at least 20 dB below the maximum PNL of the aeroplane.

Aeroplane sound pressure levels within the 10 dB-down

Figure 6.35. The background noise problem – ambient and system noise inter-fering with the aircraft sound signal.

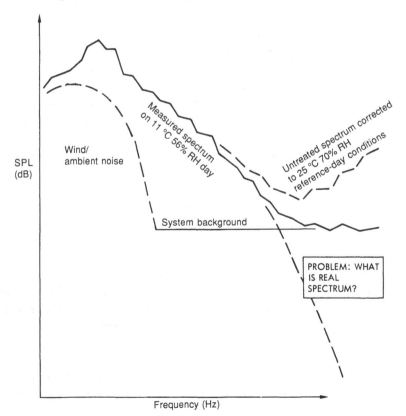

points shall exceed the mean ambient noise levels determined above by at least 3 dB in each $\frac{1}{3}$-octave band or be adjusted using a method similar to that shown in Attachment G. Where more than seven consecutive $\frac{1}{3}$-octaves are within 3 dB of the ambient noise levels a time frequency interpolation of the noise data shall be performed using a procedure such as described in 2.2 of Attachment G or by other such equivalent procedure approved by the certificating authority.

Attachment G then explains how to make an "official guess" as to what the aircraft noise level might have been if the background had not been quite so high. This is a time-consuming and highly dubious process that makes a mockery of determining whether an aircraft is too noisy or not. The system has simply grown to fill the available administrative time.

On the other hand, it is the author's belief that agreement on an international standard background level could put the problem to bed once and for all. The concept of a standard background level is not new, but it is untried. It would probably be equally useful in the context of noise exposure contours as it would to the certification process. Low-level contours spread wide across the community, where local ambient levels may already be the dominant source. Moreover, most exposure methods contain a sizable element to account for the frequency of aircraft movements into and out of the airport; often, this can amount to ten times the logarithm of the number of movements being added to the actual noise level. Therefore, in principle, an aircraft producing less than the genuine background noise could, if it operated in sufficient numbers, contribute significantly to overall exposure – even though it were never heard!

By analogy, consider the carefree bumble-bee. Flying lazily through the garden, it is audible above the local ambient, probably producing a trivial 60 EPNdB or so in a 10–15-sec flypast. A modern twin-jet aircraft overflying the same garden will be some 20 EPNdB noisier. Hence, in the passage of an hour, the innocent meanderings of a few bumble-bees could, in noise exposure terms, be equal to the take-off of a modern twin-jet. Indeed, with their "tone" correction and, if they hovered for any considerable period and extended their "duration", they might even have a bigger noise impact!

Obviously, this is an unreal situation but, surely, the sensible solution would be to disregard any noise generated by an aircraft that fell below a modest ambient background level. We should recognise that we live surrounded by natural and man-made sounds and that an aircraft close to the everyday ambient is, in fact, virtually zero in environmental consequence. It should not be too difficult to agree on an inconsequential background level and describe its characteristics as a function of frequency. By reference to Figure 6.36, this would probably take the form of a spectrum shape that fell with increasing frequency (to reflect natural background noise caused by atmospheric turbulence and wind), until it met a horizontal "floor" (which would just exceed the electronic background noise present in good-quality recording and analysis systems). That shown in Figure 6.36, in fact, equates with a noise level of around 65 PNdB, which is significantly below the value considered relevant in contour mapping and more than 20 dB below the lowest noise certification criterion.

6.5.5 Summary

If the foregoing key items were addressed in a positive manner, many of the current technical difficulties arising from technopolitical decisions could be swept aside. Measurements could be made free from ground-reflection contamination, large atmospheric absorption corrections avoided and inconsequential tones and background noises

Figure 6.36. Suggested standard background noise level.

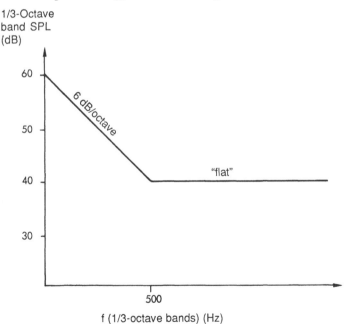

1/3-Octave band SPL (dB)

f (1/3-octave bands) (Hz)

ignored. The whole aircraft noise measurement, analysis and public presentation process would be far cleaner, more direct and representative of the environment of the "customers", the airport neighbours. Currently, they are led to believe that published levels provide a fair basis for comparison, which they do not.

7

Aircraft noise prediction

Aircraft noise prediction involves two types of activities: predicting the noise of an individual aircraft and assessing the cumulative effect of the complex pattern of operations in and out of a specific airport. The latter depends on the former for a wide range of aircraft types.

7.1   Aircraft noise

Without available direct measurements, the only method of assessing the impact of a completely new aircraft or power-plant design is to utilise a reliable prediction procedure. Such a procedure may be able to make use of a limited amount of directly relevant data, for example, engine test data where a development programme is under way, or it may have to rely entirely upon empirical component prediction procedures. The latter situation arises at the advanced project stage of any new aircraft design – a current example could be an aircraft with propfans or a second-generation supersonic transport with a novel variable-cycle propulsion system concept.

To be successful, aircraft noise prediction must be based on a reliable definition of aircraft performance and a confident prediction of the noise characteristics of the power-plant (as a function of power setting, altitude and flight speed). Where there are substantial and related measured data to support a new concept, their projection to the new aircraft situation can follow fairly well-defined routes. For example, if the new aircraft incorporates power-plants that are not much different from versions already in service, flight test data can be transposed fairly simply to a new situation. Where the measured information is obtained from static engine tests during the development programme, methods are being established for transposing these data to the flight situation.[2,6] The least predictable mode of operation embraces the totally new aeroplane concept. Under these circumstances it is necessary to rely upon accumulated past experience in the form of component-based predictive procedures.

The objective of this chapter is not to provide a total detailed prediction procedure, for no such preferred methodology exists. The purpose here is to explain the necessary elements and then to direct the reader towards the most reliable current information, via references and the summary of prediction methods presented in Table 7.1. Note that light aircraft are not included in the present discussion.

Figure 7.1 outlines the minimum elements necessary to provide a credible estimate of the noise of a given airframe–power-plant combination. The main features may be expanded as follows.

7.1.1   *Power-plant design details and performance characteristics*

At the very minimum, there should be either a design scheme for the power-plant in question and a knowledge of how the individual noise-producing component areas perform, or a credible extrapolation or interpolation of both noise and performance data from a similar power-plant. If the latter exists, then the detailed component procedures described next become unnecessary.

7.1.2   *Component noise prediction procedures*

A suite of component noise prediction procedures is required that allows all the significant noise sources to be related to leading engine performance parameters before being integrated to reflect the noise of the total system, including any reductions resulting from specific noise-control actions. The necessary depth of detail and breadth of coverage of the component procedures are related directly to the type of propulsion system, the design of aircraft in question and the aims of the prediction exercise. For example, prediction of certification-type noise levels will demand a knowledge of all the sources that lie within 10 dB of the peak level, whereas a prediction of notional levels at large distances will be controlled by the low-frequency sources, and it might be possible to ignore others.

In any prediction, it is normally the propulsion system noise that controls the situation, and this must be the principal concern. Fundamentally, there are three types of propulsion system to consider, the jet (or turbofan), the propeller and the lifting rotor of the helicopter. Let us consider firstly the jet, as the prime offender over the years, and then address the others later. Typical "jet" noise sources, as they affect the overall flyover noise–time history, are given in Figure 7.2.

With zero- or modest-bypass-ratio engines, it is necessary to have procedures covering:

Table 7.1. *Summary of noise prediction methods*

Noise source	Primary methods	Supplemental info
Engine-order tones (buzzsaw)	ESDU[945] NASA TMX-71763[962]*	
Subsonic fan	NASA TMX-71763*	Benzakein & Morgan[961]
Compressor	NASA TMX-71763*	Smith & House[964]
Turbine	NASA TMX-73566[968]*	Matta et al.[969] Mathews et al.[967] Smith & Bushell[491]*
Combustion	SAE ARP 876C[947] Appendix D	NASA TMX-71627 Ho & Doyle Emmerling et al.*
Jet (single-flow, round nozzle)	SAE ARP 876C* Appendix A	ESDU NASA TMX-71618[949]* NASA TMX-81470*
Shock-associated	SAE ARP 876C* Appendix B	NASA TMX-79-155
Jet (coaxial flows)		SAE AIR 1905[956] NASA TMX-71618 NASA TP-1301* (inverted) NASA CR-3176*
Jet suppressors		ICAO-CAN-WGE[804]
Flight effects		Bryce[954] Cocking[951] NASA TMX-79155[952]
Airframe	Fink, FAA Report RD-27-29	NASA TND 7821 NASA CR-2714
Blown flaps	NASA TMX-71768[983]	
Absorptive liners	Kershaw & House[966]	
Propellers	SAE AIR 1407[970] NASA TMX-83199[986]	
Helicopters	NASA TMX-80200[973]	
Whole aircraft (SSTs only)	NASA TM-83199[986] ICAO[20]*	
Contours	FAA-INM[994]	ICAO[990] SAE[991] ECAC[992]
Acoustic fatigue	ESDU[960]	

Note: NASA TM-83199[986] [Aircraft Noise Prediction Program (ANOPP)] covers all engine components, using references marked with an asterisk.

(a) *Jet mixing noise:* Here, spectral levels are normally required as a function of the jet (relative) velocity, its massflow and temperature.

(b) *Shock-associated noise:* This will normally provide a spectral description as a function of jet pressure ratio, for a circular

Figure 7.1. The essential elements of aircraft noise prediction.

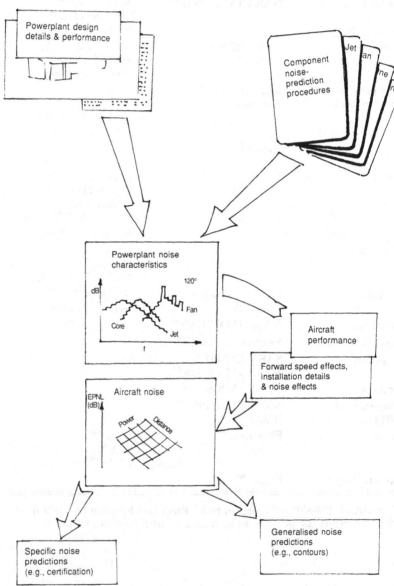

convergent nozzle. Most real engine designs would attempt to minimise the effect of shock and jet noise either by controlling shock structure via a convergent-divergent nozzle or affecting both the shock and the mixing processes with an exhaust suppressor. Hence due allowances would need to be made. The above sources dominate the noise at full engine power, but for lower power conditions on this type of engine it is necessary also to cover

(c) *Compressor noise:* Here both the tonal and broadband components are normally correlated in terms of rotor tipspeed, with corrections for massflow and rotor–stator row separation. This will allow a prediction of the noise from the engine inlet and, in the case of a bypass engine, should allow consideration of compressor noise propagated down the bypass duct.

Even so, it will also be necessary to complete the prediction of nozzle-radiated sources with a method for

(d) *Turbine noise:* The tonal and broadband content is most likely to be related to the tipspeed of the rotating blades, with corrections for massflow and the spacing between the fixed and rotating stages.

In the case of the turbofan, the source balance is somewhat different from the "jet". Fan noise often dominates in all conditions, although jet noise is still important at high powers and the combustion system may also be relevant. Hence, as well as methods covering compressor and turbine sources, the turbofan also requires

Figure 7.2. The sources of relevance in a flyover-noise–time history.

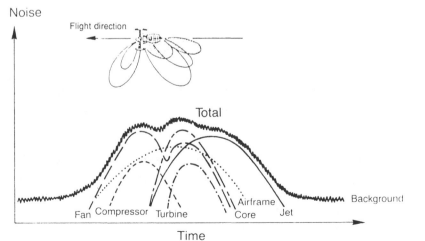

(e) *Low-velocity/coaxial jet mixing noise:* This will usually be based upon primary jet velocity but will have corrections for primary and secondary velocity and massflow ratios.

Also required is a method for

(f) *Combustion noise*

and, most important, a comprehensive prediction procedure for

(g) *Fan noise:* This must define both tonal and broadband content as radiated forwards through the intake and rearwards through the bypass duct. The method will also have to consider "buzz" tones when the bladespeed is supersonic and also cover the general question of any interaction with the core compressor.

With respect to the total power-plant installation, all internal source levels will need to be adjusted to take account of control measures such as the use of acoustically absorbent material in the intake and the bypass duct. Hence, it is important to have also a method of computing

(h) *Duct attenuation:* This affects fan, core compressor, turbine and combustor noise.

For all the above component sources, it is necessary to make appropriate corrections for changes in flight speed between the take-off and the approach conditions, and to allow for any amplification and/or shielding effects that the aircraft structure may have on the individual sources, or for new sources that might be created (e.g., jet–flap interactions). There are no readily available methods for computing these effects but, generally speaking, engines mounted under the wing will experience amplification effects whereas those mounted to the rear of the fuselage will have the benefit of shielding effects both beneath and to the side of the aircraft flight track. All these effects are normally no greater than 3 dB.

7.1.3   Aircraft noise characteristics

The above factors make it possible to predict the component spectral levels in the far-field at any given angle to the flight path. Unless it is a requirement to maintain spectral information in fine detail throughout the noise–time history of an aircraft flyover, it is normal to integrate total flyover noise energy to produce a single-number expression of the noise of a single event at a given power setting (e.g., EPNL, peak PNL, peak dBA or SEL). However, before this process can take place, it is necessary to include another relevant source:

(i) *Aircraft self-noise, or airframe noise:* This source of noise varies not only with flight speed and mass of the aeroplane, but also with the configuration. As discussed in Chapter 3, the most important features are usually the flaps and landing gear, which have their biggest effect during the landing approach. The procedure used to predict airframe noise should provide spectral information if this source is to be integrated into the total flyover level in the same way as the engine components. Otherwise, an approximate method of adding airframe noise in terms of the chosen descriptive noise unit is necessary. Either way, on most modern aircraft, airframe noise can add up to 3 EPNL to the turbofan noise on the landing approach, but less than 1 EPNL at other conditions. It will be less significant at all conditions if the power-plant is of low bypass ratio (say, no more than 1 EPNdB).

Having compiled a suite of noise prediction procedures for the power-plant (including the airframe and installation effects), these should now be allied to the aircraft performance so that a "carpet" of noise against engine power and distance can be constructed for the relevant flight speeds. For example, take-off flight speeds are usually in the region of 0.25–0.3 Mn and the approach is about 0.2 Mn. This carpet of noise, power and distance will only apply to the one-engine "over-flight" condition, and it will then be necessary to make further adjustments for other factors. For example, it is necessary to take into account the effects of having more than one engine on the aircraft, and the position of the engines. It may be that some special installation effect can be computed from previous data; or there may be some shielding of the noise by virtue of the installation. Examples of these are the interaction between the jet from a wing-mounted engine and the wing flaps, and the centre engine installation of the trijet. In the latter case, noise from the inlet is not heard beneath the aeroplane, but it becomes progressively audible as the observer moves to the side of the flight track.

So-called lateral attenuation, which often includes the absorption of sound propagating for large distances over the ground and the effects of the measurement position (ground reflection) also have to be taken into account. When these are accounted for, the complex noise signals received from the aeroplane can be integrated and expressed on an appropriate scale (EPNL, SEL etc.). When this is repeated for a range of aircraft flight conditions and engine power settings, the comprehensive noise database for a particular aircraft may then be used to provide an indication of likely certification levels. Equally, it may form the basis of the input into a contour prediction program for the purposes of evaluating noise impact at a particular airport.

7.1.4   *Propeller power-plants*

The procedures for predicting the noise from propeller-powered aircraft are identical to those for the jet, except that the components are different. On some propeller power-plants the compressor and turbine methods outlined in (c) and (d) above may be useful, but it is the propeller itself that normally dominates the situation – certainly at full power, and often at the lowest powers for the landing approach. Hence there is a fundamental need for comprehensive prediction procedure

(j) *Propeller noise:* This should take into account the tonal and broadband components, as well as any interactive effects brought about by two-row configurations, such as those being proposed for high-speed application. In addition, the method needs to take into account the effects of asymmetry in the flow (to cover the effect of aircraft attitude) and, where appropriate, the effect of changing propeller pitch. Equally, the propeller-wake interaction with the wing of the aircraft (in a conventional "tractor" installation), or the interactive effect of a wing wake or pylon mount wake entering the propeller flow to disturb the otherwise uniform inflow (in a "pusher" installation), need to be taken into account.

Unfortunately, there is only a limited amount of published information on propeller noise prediction, and generalised references have to be relied upon. Note that the noise from the exhaust (jet) of a propeller power-plant is usually of no consequence.

A noise–power–distance map of information can be constructed for a propeller-powered aircraft in the flyover mode in exactly the same way as the jet, after due consideration of any installation effects. It is the prediction of the noise to the side of the flight track that becomes somewhat more difficult, for, although ground-reflection effects are identical for all aircraft, the lateral effects with an unducted rotor are not. The noise signals propagating to either side of the aircraft are quite different.

Figure 7.3 illustrates the magnitude of the difference, often a minimum of 5 EPNdB at the peak in the noise–time history and as much as 10 EPNdB at other times. There are believed to be at least two reasons for this effect.

Figure 7.3. Observed asymmetric effects in the noise radiated to each side of a propeller-powered aircraft.

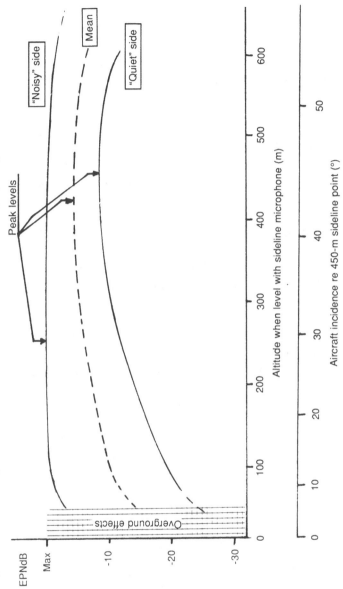

Firstly, there is a source effect for, during take-off and landing, the airflow into the propeller is seldom in the direction of the axis of the rotor. Aircraft incidence to the flight path and upwash effects over the wing can set the propeller rotor axis at considerable incidence to the free-stream flow direction (Fig. 7.4). As a consequence, when the propeller is moving upwards, it is at a different incidence to the free-stream flow than when it is moving downwards, and also has a marginally lower tipspeed. Although the side-to-side variations in aerodynamic loading are not so great, it is a situation analogous to the moving helicopter (see Chapter 3), where the source noise and radiation characteristics are quite different on the two sides of the rotor. Hence, depending on whether the propeller rotates clockwise or counterclockwise, the noise to one side of the aircraft will be higher or lower than to the other.

Secondly, and either amplifying or negating the above effect, when the engines are mounted conventionally on the aircraft wing, there is a marked side-to-side effect resulting from fuselage shielding or reflection effects. Again, the direction of the effect is a function of the direction of propeller rotation. Figure 7.5 explains the situation.

Noise generated in the high-velocity region at the tip of each propeller blade, and propagating strongly in the direction of rotation, is normally free to radiate tangentially from the rotor disc. However, in the installation shown, with each propeller rotating in a counterclockwise direction (when viewed from the front of the aircraft), noise

Figure 7.4. Asymmetric noise generation effects due to wing airflow upwash into a propeller.

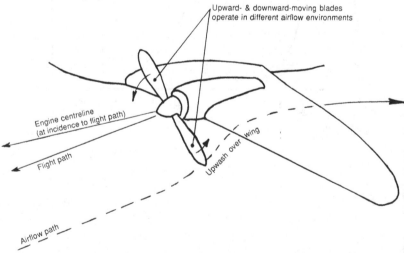

generated when the blade is moving through the top of its trajectory is free to radiate from the starboard propeller in a starboard direction, but that from the port propeller suffers physical interruption from the fuselage. Consequently, the noise level received on the port side of the aircraft will be higher than that received on the starboard side, both by virtue of the shielding of the port propeller by the fuselage and possible reflective amplification of the noise signals from the bottom of the propellers, which radiate in the port direction without interruption.

As already observed, differences in noise level as measured on either side of propeller-powered installations can be considerable. However, the differences can be eliminated if the power-plants are "handed", with for example a counterclockwise rotational direction on the port side being balanced by a clockwise rotation direction on the starboard side. In this way, both the generation and propagation effects are identical on both sides of the aeroplane, one side of the installation being a mirror image of the other. Such power-plant handing is not usual, for it demands duplication of many parts, but it can be considered a way of eliminating any extremely high noise levels to one side of an aircraft.

Another large difference between propeller and engine noise prediction lies in the ability to use data acquired statically. The aerodynamic performance of a propeller run statically, and hence the noise charac-

Figure 7.5. Shielding and reflection of propeller noise to the side of the aircraft.

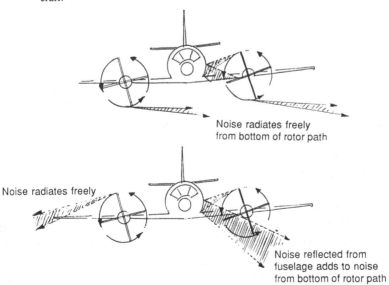

Noise radiates freely from bottom of rotor path

Noise radiates freely

Noise reflected from fuselage adds to noise from bottom of rotor path

teristics, are wholly unrepresentative of those under forward speed conditions.[610-14] Whereas some noise components on the jet engine may be transposed directly to the flight régime and others may be transposed with due allowance for generative effects and both Doppler frequency and amplification effects, this is not the situation with the propeller. Only the noise of the power source (engine) can be considered broadly to be the same in flight as on the ground.

It should also be noted that ground-reflection interference has a far greater impact on the low-frequency tones generated by a propeller than on the equivalent tones generated in the higher frequencies by the fan, compressor and turbine units of either a jet engine or a turboprop unit. The fundamental tone can be completely augmented or cancelled, as shown earlier in Figure 6.23.

7.1.5   The helicopter

In the case of the helicopter, methods developed for jet and propeller power-plants are of little relevance. The unique effects of having an "advancing" and "retreating" blade in the main rotor of a helicopter, coupled with the nonaxial inflow conditions of a tail rotor make helicopter noise prediction extremely difficult. The additional effects of blade slap, which are frequently a function of the way in which the flying machine is operated, and interactive effects between the main and tail rotors (or tandem main rotors) almost make each type of helicopter unique. Nevertheless, attempts have been made to quantify the detailed component noise sources, which can be used in building a picture of the main noise features of a helicopter,[972-4] and one attempt has been made at a generalised prediction procedure.[973] Otherwise, a helicopter is such a special vehicle that prediction of its noise characteristics is normally only possible by a specialist in that industry. At this stage, with the civil helicopter fleet being only the size of the large propeller fleet and often operated on unusual routes, the subject must be regarded as being of local, rather than widespread interest. The general provisions of noise certification will give some guidance to the likely total noise output of helicopters of varying weights, and hence mission duties.

7.2   Aircraft noise contours

The presentation of noise impact by the use of contours of constant noise level plotted on a local map is a visual way of expressing either the general situation or a change in noise "entropy". Contours have been used for many years by local government, national authorities, airports and aircraft manufacturers and, in most cases, the selection of the noise measure and the absolute level of the contours plotted has been a function of the particular objective of the exercise. For example, an aircraft manufacturer wishing to indicate the degree of improvement in a new aircraft over an older version will choose both the noise measure and the absolute level of the contours to indicate that large areas around an airport see a benefit from the operation of the new aircraft. A local authority will use the same ploy to try to demonstrate to the public that the actions it has taken have resulted in a significant improvement in the local environment, whereas a group of antinoise campaigners may use the opposite tactics, selecting a unit and a level that show minimum change.

This is not to say that everybody who uses a contour presentation is acting in an underhand or devious manner! It is natural that one should want to support in any reasonable way possible the case that is being made. A similar philosophy applies to the way in which noise exposure indices are constructed. The creator of an index will use his or her best judgment in including or excluding those elements seen as relevant or irrelevant in the particular circumstances. A local authority may well favour the use of a dBA-based index, because it bears some relevance to local statutes concerning industrial and urban noise control. A national body will probably use a scale based upon the noise certification unit, since the available database may be expressed largely in terms of that unit. A manufacturer will use a multiplicity of units − those that satisfy the needs of national and local authorities, and those that present the best case to the airline customers.

Hence, it is not surprising to find that a multitude of noise-contour calculations and presentation procedures have been developed over the years. Every nation seems to have constructed its own measure of aircraft noise exposure, and experience has shown that no two procedures solve a problem in exactly the same way. For example, the noise pattern at one European airport on the boundary between two nations was evaluated using the two different methods established by the nations concerned. Quite different impact assessments emerged, which led to different policies being proposed and the two countries to conduct an urgent appraisal of the methods used. Unfortunately, as in these cases, nationally instituted rating scales and contour computation procedures have often been recognised by being included in a statute which, if subject to change, would take several years to unscramble. Whilst this is unfortunate, on the brighter side, the kind of mismatch mentioned above has meant that the 1980s have seen a considerable effort devoted to seeking international harmonisation of both indices and contour prediction procedures. Over the coming years, these moves are likely to affect many of the predictive methods that are now established in the legal structure of major nations, and impact future airport noise-control policy.

In the United States, the noise exposure forecast (NEF) was used for many years as a tool to shape the conclusions of important environmental impact statements. Although the NEF has now been replaced by the DNL, the FAA-devised computer-based integrated noise model (INM) is still used to examine the situation at the nation's major airports. Japan and Italy have used a version of ICAO's recommended weighted equivalent continuous perceived noise level (WECPNL) to express the changing situation around their airports. In the United Kingdom, the noise and number index (NNI) has been allied to national legislation for a number of years, and officially computed NNI contours have heen published annually to indicate the changing situation around the United Kingdom's major airports. Contour area has also been used to dictate the acceptability of particular aircraft for operation at night. Many other nations have their own indices and contour calculation programmes, ten of which are listed in Appendix 1.

The one thing that all contour prediction routines have in common is that they require a credible input, both in terms of the noise of individual aircraft and their aerodynamic performance. Otherwise it is quite simply a case of "rubbish in – rubbish out". Along the way, each programme has to include assumptions about the way in which noise is generated and radiated by the aircraft, and how it propagates to the observer on the ground under the wide range of operational techniques that are used. Some of these assumptions have varied from model to model because of a lack of understanding of the physics or the use of different background databases. Many technical and political issues were highlighted in the late 1970s, and a genuine effort has been made in recent years to iron out all the differences, so as to produce a harmonised "recommended" prediction procedure.[990] The typical procedure is addressed in Appendix 5, but let us deal here with some of the major issues.

7.2.1   Technical problems

Technical problems are mainly associated with predicting the way in which noise propagates from an aircraft to any chosen position on the ground below. Generally speaking, the only noise information available to the inventor of a contour prediction program is that resulting from noise certification testing or ad hoc/airport monitoring data. Flight noise testing is extremely expensive, and "quality" tests outside the sphere of noise certification are not normally conducted unless there is some clear and specific research or development objective, and the information is rarely published at an early date. Moreover, although certification testing is a source of good data, apart from limited information to the side of the aeroplane at one fixed distance, it is concerned only with noise data acquired directly beneath the aircraft flight track. Although these tests relate to both the take-off and approach phases of operation, the data they yield are reliable for only a small percentage of the total land area affected by an aircraft operation. This is because the biggest unknowns are the way in which the noise radiation pattern is affected by aircraft manoeuvres and propagation changes as the observer moves from beneath the aircraft flight track to a position some distance to the side.

Observations over the years have suggested that there is normally an appreciable reduction in the noise received at positions to the side of the aircraft flight track compared to measurements taken beneath the aircraft. This effect is known generally as "lateral attenuation". Some contour prediction programmes use empirically based corrections for lateral attenuation effects, others merely use hypothetical adjustments. For many years, there was no consensus or referenceable documentation to provide confidence in any of the methods being used.

Accordingly, when some nations started to use noise contours in a pseudoregulatory manner (for example, in allocating night operational quotas between airlines) and others threatened to use contours in rigid certification-type structures, all concerned became interested. In particular, it focused attention on those issues that had a big effect on noise-impacted areas, like lateral attenuation. Much of this activity occurred at about the same time that moves were being made in Europe to harmonise contour methodologies, through the European Civil Aviation Conference (ECAC),[992] an organisation composed of twenty-two member states. It was in Europe that there was greatest conflict between neighbouring nations in the presentation and interpretation of noise exposure information, for Europe has nothing equivalent to the federal preemption possible by the FAA. The European Economic Community (EEC)[993] is the nearest equivalent, but it is a young organisation and, despite the fact that it is dedicated to environmental preservation, its edicts are applicable to only just over half the number of countries that subscribe to ECAC.

ECAC worked hard for about two years in trying to put together a harmonised contour methodology, during which time many of the technical problems were exposed, as well as ECAC's determination to solve them. In this, they sought help from the wider technical community, and in particular members of the U.S. Society of Automotive Engineers (SAE) who, in turn, were prompted to accelerate their own technical programmes.[991] One of these, which is vital to the successful development of any contour procedure, was the development of a credible lateral attenuation methodology. In a comparatively short time, new experimental evidence was gathered to supplement evidence that had existed for several years, and a simplified relationship between lateral attenuation, distance and angle of elevation of the aircraft was developed.[995] This procedure rapidly found its way into the emerging U.S. and European models.

Finally, in the wider interests of harmonisation, those member states of ICAO that were a party to the developing European and U.S. methodologies joined forces to encourage the development of an international methodology,[990] which was published in 1988. Thus far, nobody has offered a computer program or database that will allow the ICAO, SAE or ECAC methods to be used to model the airport situation. However, the FAA's integrated noise model[994] is freely available and, because the FAA has declared an intent to make it compatible with international thinking by 1990, it may well be used more and more on an international basis.

7.2.2   Elements of contour methodology

To be able to model the noise heard on the ground as an aircraft performs either a take-off or landing procedure, a prediction programme requires a substantial quantity of input data, along with significant manipulative capabilities. The essential input requirements are a noise–power–distance "map" and a full performance statement for the aeroplane. The noise input must, of course, be in a unit compatible with the programme, which may conduct its own time integration or may be input directly with an "integrated" unit, such as the EPNL or SEL.

The manipulative capabilities, apart from the possibility of time integration, need to include the ability to deal with changes in engine power, flight speed, flight track, distance to and angle subtended by the observer. Over and above the elements necessary to compute the noise of the aircraft whilst it is airborne and climbing from or descending into the airport, there are issues relating to the period of time that the aircraft spends close to or actually on the runway that have a significant effect on the overall size of the contour. In a typical operation (Fig. 7.6), an aircraft will be stationary at the end of the runway prior to take-off and the engines will be gradually increased to full power. In most cases these days, the brakes are released before full power is achieved, but sometimes the aircraft is held on the brakes for a few seconds whilst a systems check is conducted. Under these circumstances, there is a large lobe of noise radiated rearwards to either side of the aircraft, which is essentially constant in time until the brakes are released; but it may be necessary to compute its impact on the community in terms of a time-integrated unit. This is not easy!

Once the brakes have been released and the aircraft accelerates down the runway, the situation is more predictable, although, as yet, not as well as might be hoped.[998-9] This is because the process is complicated by a number of effects. Normally, jet noise decreases as the shear between the jet and the environment decreases with increasing aircraft speed. With a high-bypass-ratio turbofan or propeller, the effect is not so marked, although the aerodynamic conditions are changing substantially and the noise from the fan may also decrease. Moreover, during the whole of the ground-roll process and for the early part of the climb after take-off, the noise radiated by all aircraft types undergoes considerable reduction due to the effects of over-ground attenuation.[232-5] Equally, there may be some shielding of noise from an engine on the remote side of the fuselage, and so the noise pattern radiated can be extremely complex.

Once the aircraft has left the ground and begins to climb, it comes out of the effects of shielding and overground attenuation progressively and, at an altitude in the region of 300 m, the noise reaches its peak level and is free to radiate to the community virtually unaffected by installation or local topographical features. From this point on, the calculation of the noise level on the ground usually becomes compara-

Figure 7.6. A typical noise footprint.

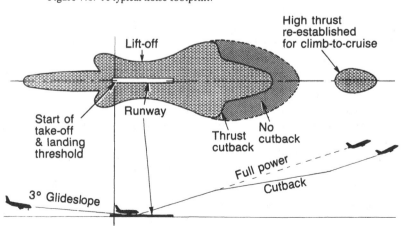

tively simple, apart from allowances for changes in engine power, aircraft configuration, flight speed, flight track and distance, already referred to. Other elements to consider include the use of power cut-back as a means of reducing noise close to the airport. As Figure 7.6 shows, this has the effect of slimming the contour locally, but when high power is later restored for the climb-to-cruise altitude, engine noise rises and an "island" of increased noise is created.

From the foregoing, it is not surprising that many computational procedures have been developed, and that different authors have different opinions on how each element should be tackled. Frequently, there are differences in both the source and the interpretation of technical information that provides quantification of the various physical effects. The moves towards harmonisation by ICAO and other international organisations are an attempt to bridge the gap between the many authors and interpretations of technical information from

Table 7.2. *Summary of differences between SAE, ECAC and ICAO contour methodologies*

Item	SAE	ECAC	ICAO
Cumulative noise measure	Day–night level (DNL)	$L_{eq}(A)$ preferred	Most OK
Input noise descriptor	Sound-exposure level (SEL)	SEL and peak dBA	SEL and peak dBA
Reference atmosphere	Identical, being set in terms of a given attenuation rate, not ARP 866, % RH & T		
Noise/power/distance data source	Certification (layered and unlayered) Other tests Predictions	Certification (layered) Predictions	Certification (layered and unlayered)
Aircraft performance input	Equations supplied (coefficients are not supplied)	Prescribed flight profiles, with equations to give variants	
Reference conditions	Identically framed for atmosphere, wind, runway altitude, aircraft mass and engine power		
1. Lateral attenuation	SAE AIR 1746 used in all cases		
2. Take-off ground roll	Same polynomial directivity used in all cases		
3. In-flight turns	Outline method	Detailed method	Example method
4. Flight path dispersion	Need noted only	Data given to allow calculation of effect	
5. Noise grouping	Indicated possible	Detailed guidance	Indicated possible

various sources. Even so, there are still some differences between the major methods, the most important of which appear in Table 7.2.

7.2.3 Validation and accuracy of contours

It must be recognised that contours are extremely sensitive to small changes in input sound level. For example, it can be shown easily that a change of some 4–5 dB in the level of source noise will effect a doubling (or halving) of the area of ground enclosed by a given contour of constant level. If, therefore, the accuracy of the input noise data is only, say, ±1 dB, then the contour area could vary by as much as 40%. This is no insignificant change and is the fundamental reason why single-event contour areas, or "footprints", have not been readily seized upon as a methodology for certification, or used widely as an airport noise control.

Because of this sensitivity to assumed source level, contour prediction procedures are widely recommended for only one purpose, the long-term indication of the variation in noise exposure around a given airport resulting from changes in the operational pattern and/or aircraft fleet mix. Contours embracing large areas of land can only be regarded as a planning guide, since their reliability is inversely proportional to distance from the airport. The greater the distance from the airport, the greater the variability due to overground or "lateral" attenuation effects and changes in atmospheric absorption resulting from the ever-varying conditions of temperature, humidity, wind and turbulence.

It is really only the contours reflecting the higher levels, that is, those closest to the aircraft flight track, that are reliable enough to form the basis for absolute assessments, or that might be related to the "published" certification levels of any given aircraft type. For example, the width and length of a close-in take-off contour and the length of an approach contour could be calculated from the levels at the three certification measuring positions (see Section 8.5).

Despite the general sensitivity of contour area to accuracy of input, "validation" of methodologies has been attempted, and average monitoring data have been within ±2 dB of predictions in many cases.[991] This gives a good measure of credibility of the input in what is an extremely complex systems analysis process, but it still means that the long-term average contour areas can be predicted in error by up to 80%. Hence, they can really only be used to estimate any change to the annoyance level likely at a particular point in the community, and not to indicate the absolute level of noise at any given point.

8

Prospects for the future

Over the years, much has been said about the future. Usually, environmentalists predict that worse is to come, the industry expects a rosy future, whilst governments try to present a "balanced" view.

The future always depends on the starting point; and no two observers, or victims of aircraft noise, have the same viewpoint. Some live near well-developed and possibly operationally saturated airports and therefore experience the benefits of advancing technology in the form of lower individual aircraft noise levels and a generally improving environment; others are adjacent to new or expanding fields, where the growth in the number of operations is the overriding factor; a small percentage of people would complain in any case – if aircraft were silent there would be some who would perceive a "stealth" factor.

The importance of aircraft noise in society in the future will be a function of several factors. Uppermost are whether the manufacturing industry will be successful in advancing noise control technology, whether air transportation is going to expand to any significant degree and how sensibly government and airport operators apply controls. All these factors bear examination in trying to build a picture of the future, as does the question of whether the available noise-forecasting tools are right for the job. This is most important; the weighting applied to individual elements of forecasting methods must be examined, since they can have a tremendous influence on predicted trends in noise "exposure" and, hence, any decisions based on those predictions.

The main elements that bear examination, probably in order of importance, are as follows:

> *The status of technology,* that is, the absolute noise level imposed on the public by each aircraft movement, including

duration and spectral abnormality factors (tones, whines, screeches, rotor slap, propeller beating, etc.).

The operational pattern at the airport, that is, the frequency of aircraft movements over critical communities, and hence the degree of intrusion on life.

The local environment, in particular the normal ambient noise level – which is always lower at night, and hence makes aircraft seem more annoying.

The economic situation, whether it is the controlling factor in the health of the air transport system or whether it is only important locally in terms of the prosperity of the airport and the community that relies upon it for its wealth.

Each of the above elements has to be considered carefully, for absolute noise levels are a function of advancing technology and airline decisions on which aircraft to buy; frequency is a function of economic activity and fleet mix; the ambient level varies not only with time of day, but also locality. As regards economic factors, who can predict the world or national economic situation in the medium-term future? Therefore, let us discuss those elements that are predictable with some confidence and then proceed to develop a general forecast of the future situation.

8.1 Fleet size and composition

The steady growth of the free-world commercial jet fleet has already been catalogued at the start of this book, in Figure 1.1. From its starting point in the 1950s, the number of jet aircraft has grown to over 8000 at an average rate of expansion of some 4% per annum. This growth has included the development of larger aircraft, to satisfy a continuing growth in demand for "available seat-miles" (ASMs), the industry's capacity indicator, of around 5% per annum since the mid-1970s (Fig. 8.1). By comparison, the fleet of major propeller-powered aircraft has remained almost constant in number, at around 2000, making the total commercial fleet of the order of 10000 aircraft. Although there are tentative plans for the reemergence of propeller power-plants of an advanced nature, all the aircraft currently being delivered or on offer for sale to the airlines with a seating capacity approaching a hundred or more are jet powered.

From projections of capacity requirements through to the next century (Fig. 8.2), which assume world economic and political stability, the likely increase in demand for seats is expected to be at least 4%

per annum. Even with due allowances for the steady increase in the size of each new aircraft development, the growth in fleet numbers is likely to be some 2–3% per annum, a rate of expansion that will lead to a jet-alone total of over 10 000 at the turn of the century.[46] On the basis of number of movements alone, this growth would be equal to an increase in noise exposure of over 1 dB globally. Fortunately, however, individual aircraft noise levels have been reduced considerably, and still tend to fall year by year – not by very much, but, as Figure 8.3

Figure 8.1. Historic growth of total free-world aircraft capacity in available seat-miles (ASMs).

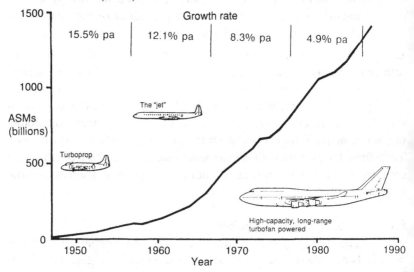

Figure 8.2. Forecast growth of ASMs by aircraft size.

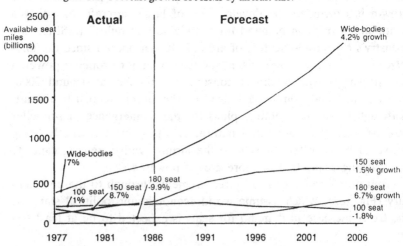

Figure 8.3. Progress of aircraft noise control.

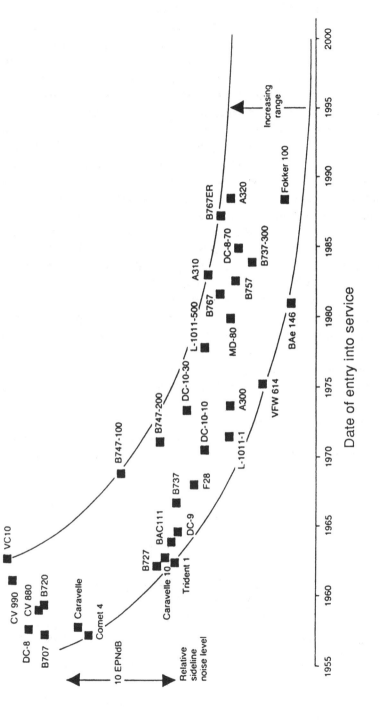

shows, the trend is still downwards. Moreover, just as important is the way in which the mix of aircraft types in the fleet is changing.

Although the fleet has grown by over 1000 aircraft in the past five years, the number of three- and four-engined aircraft has actually fallen. This is because the twin is no longer merely the traditional small-capacity short-range workhorse. Many twins now carry twice as many passengers as the older long-range four-engined types, such as the B707 and DC8 and, moreover, in late 1983, Air Canada broke new ground when it used a twin-engined aircraft on the transatlantic route – historically, a "hazardous" long-range operation and hitherto the sole province of aircraft with at least three engines. Since then, there has been a significant change in attitude towards the twin, which has allowed many long-distance routes to be opened up to this type of aircraft. The previous restrictions, which meant that a twin always had to be no more than sixty minutes away from a diversion airfield in case an engine failed, are being relaxed to reflect the fact that modern aircraft engines are an order of magnitude more reliable than they were in the days of the piston engine, when the rules were laid down, and that it is aircraft systems rather than engine reliability that dictate how long an aircraft may safely stay aloft. As a result, the airlines are now able to consider the option of using "extended-range" (ER) twin-engined aircraft over a much greater proportion of their route network, without flying much further than aircraft with three or four engines, as would normally be the case in such operations as the transatlantic sector (see Fig. 8.4). Moreover, in turn, the ER-twin means fewer engines in the maintenance circuit, with an attendant lower spare-parts inventory, and some savings in fuel consumption as a

Figure 8.4. Different transatlantic crossing routes for 60- and 120-minute single-engine diversions.

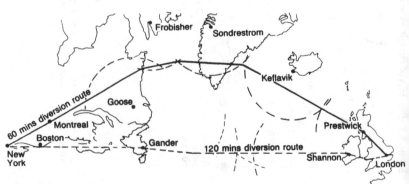

result of improvements in the aerodynamics of a "cleaner" wing and reduced structure weight.

Coincidentally, the airport neighbour should be pleased for, as Table 8.1 shows, the twin is quieter than the trijet, which in turn is quieter than the "quadripod". This is in part due to the fact that the twins listed are lighter than the other aircraft, but it is not the main factor; nor is it because the twin has the least number of engines, for the thrust required to keep a given size of airframe airborne is unchanged. Much of the improved noise comes as a result of the twin climbing away from the airport much faster than the others, thus putting a much greater distance between itself and the community. Like the range restrictions applied to the twin, this higher climb rate results from the airworthiness (safety) requirements, which demand that an aircraft be capable of becoming airborne even if it experiences a one-engine failure during the most critical part of the take-off run. This means that a twin has 100% more thrust than it needs to get off the ground, a trijet 50% more, but a four-engined type only 33%. The twin is analogous to the high-performance sports car when compared to the family "coach and four".

The overall impact of the predicted increased demand for capacity allied to the changing "shape" of the world fleet is such that some 70–80% of aircraft could be twins by the early years of the next century, compared with around 40% today (see Fig. 8.5). Because of the trend towards twins, individual aircraft noise levels in the commu-

Table 8.1. *Noise levels of modern turbofan-powered aircraft*

	Perceived noise – EPNL	
	Take-off	Approach
Four-engined		
Boeing 747	104	106
Three-engined		
Douglas DC10	98	106
Lockheed TriStar	98	102
Two-engined		
Boeing 767	89	102
Airbus A300	90	102
Boeing 757	86	98
Airbus A320	85	92
Fokker 100	84	93

nity will continue to fall without any further advances in noise-control technology.

8.2   Technology

The 1970s heralded a step-change in the technology of engine noise control, and a new breed of much quieter aircraft. This was aided by improvements in the aerodynamics of the airframe, which resulted in less thrust being required to lift each kilogram of aircraft weight. As previously discussed, that step-change was a side-effect of the move to the high-bypass-ratio turbofan, made possible by advances in materials and core-engine cooling technology. No second such step-change has yet emerged, nor is it foreseen. In fact, the much vaunted fuel-efficient advanced "propfan" is still in the gestation period and, while its future viability will depend heavily upon fuel price, other important factors include its attendant and new noise problems.

As to what can nowadays be described as a "conventional" turbo-fan, one design-cycle change that may occur in the cause of improved fuel efficiency is the growth of the single-stage fan to a bypass ratio in the 10–15 region, with an attendant reduction in fan-pressure ratio. The commercial success of such an engine cycle will depend upon whether higher component efficiencies can be achieved and the overall aerodynamic drag of the power-plant can be minimised. Just what the "optimum" cycle is from the point of view of thermodynamic and propulsive efficiencies, is a moving target. As component efficiencies take the benefits from research and development and as core engine

Figure 8.5. Projected growth of the world fleet.

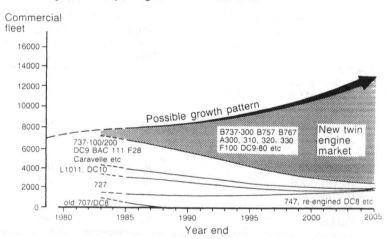

temperatures rise, it will become possible to design for improved propulsive efficiency. Figure 8.6 amplifies this point.

At any point in time, the optimum bypass ratio or, more correctly, the optimum specific thrust (thrust per unit mass of air used) for a given installation is a function of several factors over and above the "performance" of the engine in its broadest sense, before installation

Figure 8.6. Optimising a power-plant design for performance and noise.

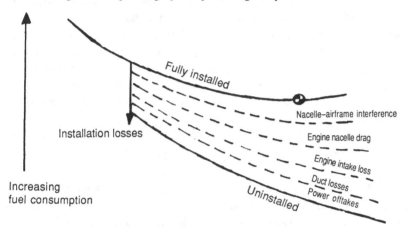

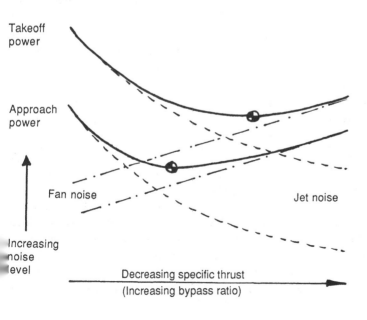

on the aircraft. As the upper half of Figure 8.6 shows, reducing fuel consumption goes hand in hand with reducing specific thrust of the engine, the optimum being dictated by the level of component efficiencies available. However, the complex process of creating a complete power-plant, including the manner of its installation onto an airframe, produces a series of penalties against the basic performance of the "bare" engine. Firstly, the aircraft requires air and power offtakes from the engine to provide essential aircraft and passenger life-support systems. These are not a function of the performance of the engine or the way it is installed on the aircraft, but they are a function of the aircraft size and range. They are demands that would be placed on any power-plant, irrespective of its design philosophy. However, there are several performance penalties that are related to the power-plant.

Firstly, there are the losses that are a function of the particular engine design and its "isolated" features. These include aerodynamic effects within the air and gasflow passages (pressure and drag losses in the intake, bypass and core-engine exhaust ducts) and the external drag resulting from airflow over the nacelle. In addition, there are the "installation" losses associated with the particular location of the engine on the airframe. These can have a major effect on final performance, in that the power-plant has to be able to accommodate local airflow patterns – upwash over the wing being one. Also, there are significant interference effects between the power-plant nacelle and the adjacent aircraft surface, whether it be the wing or the fuselage (depending on where the engine is mounted). All these losses are a unique function of the particular installation – for example, a side-of-fuselage or centre-engine installation in the tail of a trijet can produce greater losses than the freely suspended mount from a wing.

Hence the final "fully installed" performance of a particular power-plant in a particular airframe is unique to that combination. What is common to all installations is the fact that any apparent improvement in fuel consumption in the bare-engine design, contingent upon reducing specific thrust, becomes eroded as all the installation losses are acounted. It follows that the lower the specific thrust, and hence the higher the diameter of the engine for a particular overall thrust, the greater the drag losses associated with the larger surfaces. Eventually these overcome the benefits arising from reducing specific thrust. The current "fully installed" optimum for an all-new engine design stands at a specific thrust of around 5–10, or a bypass ratio of roughly 10:1. This can be compared with the situation twenty-five years ago, when the optimum bypass ratio was around 2:1; and fifteen years ago, when turbine cooling and combustor technology had allowed higher temperatures to be accommodated, and the JT9D, RB211 and CF6 "big fan" engines emerged with bypass ratios around 5 : 1. Undoubtedly, in another fifteen years, the optimum may well be of the order of 20. Technology does not stand still for long.

Nowadays, the dilemma that designers face when they are about to freeze the details of a new power-plant is not limited to considerations of overall performance. Noise has a considerable impact. Shown in the bottom of Figure 8.6 are the equivalent relationships between noise and specific thrust for the two important operational conditions of high power (for take-off) and low power (for approach). At low power, jet noise is of little significance but, because of the importance of noise from the turbomachine, the optimum point for minimum approach noise is at a comparatively high specific thrust (or low bypass ratio). It is for this reason that the low-bypass-ratio JT8D-200 series engine makes the MD 80 one of the quietest aircraft on approach. But, take-off is another story for, at high power, jet noise is more important and there is a delicate balance between the returns from reducing specific thrust and reducing jet noise whilst at the same time increasing noise from the bigger turbomachine. What is certain is that the optimum noise design point is never the same for the take-off and the approach conditions and that, invariably, neither of these two optima is coincident with the performance optimum. The designer, therefore, has to be very careful in choosing the specific thrust that will give the greatest performance benefit but still allow the aircraft freedom of operation at noise-sensitive airports.

In noise terms, the modern turbofan is an extremely complex machine. The tonal and broadband noise resulting from interactions in the early stages of compression, and in the turbine unit, combine with noise from the combustor and jet mixing process to produce a multitude of sources of almost equal importance. Gone are the days when a single source (jet mixing noise) dominated the situation. Taken across the operational power range, there are now a minimum of eight engine sources to consider, plus the airframe self-generated noise, which appear in different orders of priority depending upon the flight condition.

Research is not only producing smaller returns, but is becoming more expensive as bigger and more complicated turbomachines are required to conduct meaningful experimental work. The added complication of new sources associated with the open rotor only serves to emphasise that we are close to a technology "floor". The designer is finding it increasingly difficult to absorb the benefits of improved component performance alongside associated needs for shorter and shorter engine ducting, without causing sufficient noise to be generated to bring the final power-plant/airframe combination into violation with noise certification and local airport noise requirements.

If one considers the rate of progress over the past twenty-five years, as exemplified in Figure 8.3, it is difficult to be optimistic about noise improvements in the future. If technology progresses at the same rate (for the same engine cycle) as has been achieved over the past fifteen years, during the era of the turbofan, the remainder of this century is likely to see only about a 1–2-dB improvement in the noise generated at full engine thrust. Moreover, if bypass ratios do move beyond a value of about 10, then there will be a detrimental effect in that it will be more difficult to use sound-absorbent materials in engine ducts, for they will be shorter (relatively) with respect to engine diameter. This is particularly important at powers less than the maximum thrust used for the early stages of take-off, because the noise trend against bypass ratio is already upwards rather than downwards at approach power, and could soon follow the same pattern higher up the power range.

Designs will move forward,[37–43] driven by the necessity of maintaining the historic trend of reducing fuel consumption to create additional range or capacity, or simply to reduce the cost of every seat-mile available (Fig. 8.7). Unfortunately, the noise trend against bypass ratio

Figure 8.7. Typical improving trend in fuel efficiency – long-range four-engined aircraft example.

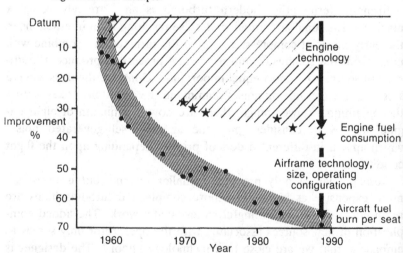

may be upwards at all power conditions. The open-rotor concept, or advanced propeller, with an effective bypass ratio of the order of tens rather than units, has to have all its noise control designed into the basic turbomachine. By definition, an "open" rotor has no encircling duct in which secondary noise control may be achieved by the use of absorbent material. With a single-stage propeller this is not too difficult, for there are known techniques for minimising the major sources of a given design configuration; but with modern high-speed designs, where there are two counterrotating rows of propeller blades, each interacting with the other, the problem is not straightforward. Propfans have been flying in demonstrator form since 1986, the most well-known type being a counterrotating two-stage "pusher" device, as shown in Figure 8.8. In its early flying, one demonstrator power-plant had reported problems with blade excitation, and another with cabin noise at cruise, and both investigated how much of a limitation noise was likely to be with such concepts.

Figure 8.8. The GE-UDF (unducted fan) on a McDonnell Douglas flight demonstration vehicle.

Surprisingly, however, this design constraint is not necessarily associated with noise as perceived by the communities around the airport, but as that transmitted to the passenger cabin during cruise operation. During take-off and landing, flight speeds are low and, providing the tipspeed of the open rotor is subsonic, the limitations are in the interactive process between blade rows, and the distribution of the lift forces over the surface of each blade. However, because cruise speeds of around 0.8 M_n have to be achieved to be competitive with jet-powered aircraft over longer ranges, "helical" blade tipspeeds are bound to be supersonic. Shock patterns radiate from the blades, virtually unattenuated over the short distance between the power-plant and the fuselage, to excite the local structure and cause cabin noise levels far in excess of those currently accepted by the fare-paying public, even over short distances. This problem can be solved in one of two ways, or by a combination of both, that is, through design control in the definition of the blade geometry and noise suppression between the outer surface of the fuselage and the cosmetic inner surface of the passenger cabin. Both are penalties; the former slaps a direct cost on performance whilst the latter uses up valuable pay-load or ability to carry fuel-weight for long-range operations. Both affect the overall mission capability of the aircraft.

It is clear that, if such designs are to become a production reality, any performance penalty for reducing cabin noise can only be allowed to be a small proportion of the potential savings in fuel consumption that arise from the open-rotor design cycle. Equally, other technical issues have to be addressed; for example, the extremely swept open-rotor blade has to be constructed in a manner that guards against any possibility of failure and consequent risk to adjoining structures, including the other power-plants. At the time this book was being written, it was not clear whether the open-rotor concept could emerge much before the mid-1990s as a competitive force in the air transport business. There are also indications that the fuel consumption gains may be much smaller than anticipated as ducted power-plant technology advances year on year.

Opinions on the future of ducted power-plants are likely to follow two general paths: Some, as already mentioned, will favour the use of lower-pressure ratios in large single-stage front fans, as per current high-bypass engines but with the added complication of a gearbox. Others will favour a counterrotating two-stage fan driven by an independent free power turbine mounted on the rear of the engine assembly, or at the front of the engine, as is the case today (Fig. 8.9). At the time of writing, no prototype vehicle has yet appeared to justify the savings in fuel consumption claimed for any of these concepts, which are often referred to as ultra-high-bypass ratios or UHBRs, but a good deal of effort is being put into the research programmes that are necessary to acquire the technology from which to go forward to development.

Whichever way they progress, if UHBRs are to offer appreciable

Figure 8.9. Two counterrotating ducted fan concepts.

The Contrafan

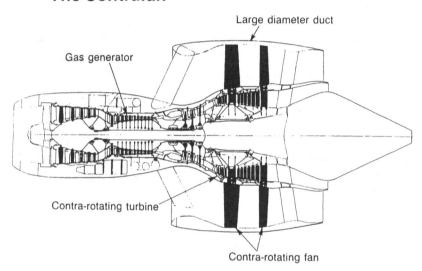

The Front Contrafan

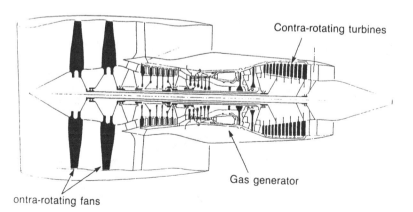

improvements in fuel consumption, their large-diameter fan assemblies will have to be "cowled" with minimum drag loss. To achieve this, the nacelles are likely to be short, thin lipped and lightweight. They will undoubtedly reduce, by a significant margin, the area available for the application of acoustic linings. This, in itself, may mean that the noise will not be containable to standards currently demonstrated by high-bypass-ratio technology, for both the UHBRs and the prop-fans lie beyond the range of experience. As Figure 8.10 indicates, the noise factor may delay their emergence for several years, for they must be as quiet as contemporary turbofans to be a realistic commercial proposition.

On balance, therefore, one can postulate that there will be little change in the average noise level of the modern turbofan-powered commercial fleet through to the early years of the next century. There may well be some improvement in the application of noise-control technology to conventional turbofans, but emerging new and radically different designs will pose problems that the manufacturers will do well to solve to anything like the degree of success on the turbofan. It would be imprudent to assume any improvements in technology will result in a continuing downward trend in overall noise from all indi-vidual aircraft types. At some point there has to be a technology floor, possibly the self-noise of the airframe, and certainly there is no indica-

Figure 8.10. Noise source variation with bypass ratio – the ultra-high-bypass-ratio (UHBR) problem.

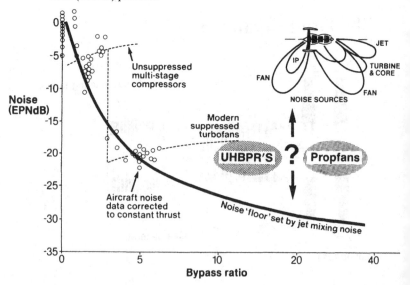

tion yet of a step-change in improvement equivalent to that which resulted from the introduction of the high-bypass turbofan.

8.3 Future noise exposure trends

At the start of this chapter, four issues were cited as being important in building a picture of the trend in noise exposure. With respect to two of them, the future shape and size of the world aircraft fleet and its associated noise output, we are able to make a judgment from previous discussions. The two issues not addressed thus far are the impact of time of day or, more correctly, intrusiveness of any external noise source above the local ambient level, and the impact of the economic climate on human reaction. Fortunately, both of these factors can be ignored in developing a generalised picture of noise exposure, for they are merely perturbations, albeit significant in some cases, in the general flow of events. Night usually follows day; and who would stake his or her life on a prediction of even the medium-term world economic climate?

To look at noise exposure in an ideal fashion, one has to select individual airports that are representative of the various types that exist around the world and examine the impact of changing the aircraft "mix" and the detailed operational pattern. Clearly, this is an enormous task and, although it has been attempted in detail for a number of the world's most significant airports, the conclusions therefrom change with time as public attitudes and "fashions" change, both in terms of exposure units and modelling techniques. It is possible, however, to make a broad judgment of the general situation by considering the way in which the world fleet is changing, that is, by inspecting trends in the number of aircraft available to ply their trade and the changing engine design philosophy – the factor that controls the noise signature that serves as a particular aircraft's trademark. A simplified scenario takes the form of that illustrated in Table 8.2. In this table, a "snapshot" has been taken of the world fleet at a particular point in time. To minimise complexity, the snapshot has been focused against a background of six critical aircraft groups – two-, three- and four-engined types, each at old and new technology standards. In this context, old and new technologies refer to low- and high-bypass engines, respectively, and the resultant six groups represent distinct noise characteristics.

For each of these categories of aircraft, certain assumptions have been made about the average number of times that the aircraft will operate each day. No distinction has been drawn between the number of operations that old and new technology two- and four-engined aircraft will perform, but there is a marginal difference in the case of the trijet. This is because many of the modern trijets are long-range aircraft, whereas the previous generation of trijets were largely medium-range aircraft. This means that the older trijets are likely to operate more frequently than their modern counterpart.

As far as noise level is concerned, the selection of the six categories was partly predicated on this issue since, if one consults the published listings of certificated noise levels (e.g., Appendix 6), it is possible to attribute an average value to each group. The subsequent application of a $10 \log_{10} N$ factor to account for the number of movements, allows a "world fleet" noise value to be calculated for each category, which can then be integrated to give a total "world exposure" level. This analysis, if conducted annually, provides an indication of the general trend in aircraft noise exposure. Let us try to "predict" the past.

Figure 8.11 plots the predicted trend from annual analyses against the left-hand scale and, as an indicator of validity, also shows the relative contour areas of the 35-, 45-, and 55-NNI values for London–Heathrow Airport on the right-hand scale. Although Heathrow is a large international airport, many of the so-called international flights cover distances no greater than those covered by domestic operations in mainland Europe or other major areas, such as the United States. For example, London to Ireland and London to France are classified as international routes, despite the fact that they only involve short-distance operations. Hence, the mix of aircraft using Heathrow, in being a broad cross-section of short-, medium- and long-range types, is probably typical of the world fleet composition, and the noise trends at Heathrow might be expected to fall in line with a world fleet projection.

Table 8.2. *Noise exposure methodology*

Aircraft type	Numbers in service (N)	Opera- tions per day (θ)	$10 \log_{10}$ ($N \times \theta$)	Typical take-off noise (EPNL)	EPNL + $10 \log_{10}$ ($N\theta$)
Old four-engined	700	2	31.5	114	145.5
New four-engined	550	2	30.5	107	137.5
Old trijets	1700	4	35.5	102	137.5
New trijets	580	3	30.5	98	128.5
Old twins	2250	6	36.5	98	134.5
New twins	425	6	29.5	90	119.5
				World fleet exposure level	147.5

It is comforting to note that, since the early 1970s when NNI values were first published, there has been a basic trend at the 35- and 45-NNI levels very similar to that projected on the simplistic world fleet basis. The 35- and 45-NNI levels are experienced by large sections of the public near Heathrow; the 55-NNI level is limited to just outside the airport boundary and is heavily influenced by the noise just before landing and the subsequent use of thrust-reverse, as well as by noise before the start of take-off and during the early part of the take-off ground-roll. Hence, the lesser decline of the 55-NNI area can be explained and is of lesser significance. Also, the perturbations in the annual trend at the lower NNI values can be attributed to the under-utilisation of available aircraft in the recession of the late 1970s coupled with the increase in economic activity in the mid-1980s. Recent years have also been influenced by the continued use of some extremely noisy old aircraft until they were "outlawed" in Europe at the end of 1986. Moreover, NNI values more generally tend to be biased upwards in comparison with other exposure metrics as a result of the use of a $15 \log_{10} N$ factor for number of operations (N) in the NNI, rather than the $10 \log_{10} N$ of the projection. Overall, therefore, we can be fairly pleased with the performance of the simple model. Having said that, any argument over past events is probably academic, since it is what the future projection says that is important in judging whether or not there is going to be a continued noise problem. Figure 8.12 looks to the future by developing the projection of Figure 8.11 through to the early years of the next century.

Figure 8.11. Comparison of projected aircraft noise trends with published noise and number index (NNI) contour areas around London–Heathrow Airport.

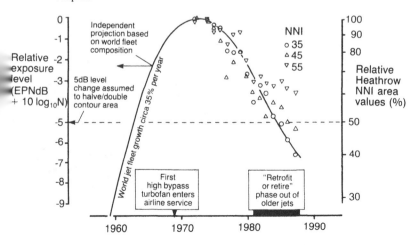

Surprisingly, even though older aircraft continue to be replaced by their much quieter modern counterparts, the trend for an improved noise climate is brought to a halt before the end of this century. This depressing and possibly erroneous forecast results, quite simply, from the importance attributed to the frequency of operation overriding the noise reductions accruing from the continuing application of high-bypass technology. Although the vast new twin-engine aircraft market indicated earlier will embody the best in turbofan engine noise control, sheer numbers override – at least, that is, if we assume that it is appropriate to continue to apply a $10\log_{10}N$ correction for the number of movements.

This point does need debating, for it is extremely important in that it greatly affects the overall conclusions about noise trends. For example, Figure 8.13 shows the impact of four different, yet quite feasible, scenarios. Firstly, we can either choose to include or ignore the remaining old noisy B707 and DC8 types that are still numbered amongst the total world fleet population, since they virtually never appear at major airports and their noise problem is reserved for "underdeveloped" areas. Secondly, we could reduce the significance of the number of operations as aircraft become quieter and this, combined with the B707/DC8 issue, would provide a range of trends that would allow conclusions to be drawn, ranging from "do nothing; everything is going to get better for the next twenty years" to "plan for an upturn in airport annoyance before the end of the century".

Figure 8.12. Projected noise exposure trends.

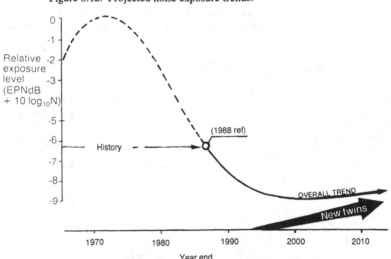

Such a dichotomy merely indicates how sensitive exposure indices are to changes in "opinion" and shows that they cannot always be relied upon entirely in making decisions. For example, one cannot allow future research and development plans to be dependent upon suspect judgments made in the psychoacoustic field, for it takes far longer to apply the beneficial findings of noise research than it takes to alter the ingredients of an exposure model!

Nevertheless, if we look at the facts surrounding the attribution of different levels of importance to the "intrusion" element of frequency of operations, we can support the more optimistic trend for the future. The reason for this, expressed diagrammatically in Figure 8.14, is that the importance attributed to the frequency factor must vary according to absolute noise level, which, in turn, relates to annoyance level, depending upon the time of day. The fact that the NNI was devised in the 1960s and most other indices in the 1970s suggests that we should be looking to a value of less than $10\log_{10} N$ as the fleet becomes even quieter in the 1990s, and that we should be using different values in different communities and at different times of the day because the mean local noise levels are different.

Put simply, the $15\log_{10} N$ factor used in the NNI has always been unrealistic since the metric was developed in an era when jet aircraft were both novel and extremely noisy and, as has been explained in

Figure 8.13. The importance of the number of operations in noise exposure evaluation.

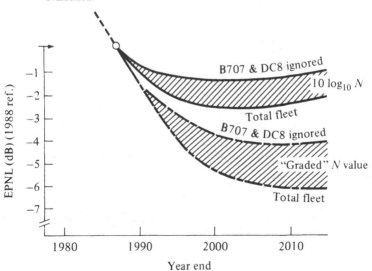

Chapter 1, "number" was not uncorrelated with absolute level in the baseline social surveys. Other noise indices, developed subsequently in the United States and in areas where air travel is a common feature of everyday life, utilise the lower value of $10\log_{10} N$. Even so, these later metrics were developed from data acquired at a time when the fleet was still dominated by fairly noisy low-bypass-ratio-powered aircraft. In other words they were, and still often are, considerably noisier than the aircraft that are now progressively replacing them. Nobody has yet experienced an "all-turbofan" environment – even the most recent study in the United Kingdom,[152] which finally recommended that a value of 9 or $10\log_{10} N$ should be used in the NNI, was constructed around surveys of reactions to operations of the 1960s and 1970s, which were nearly all old-technology aircraft. By the time the higher in-service noise levels fall further – by as much as 10 dB at a given aircraft size – it is more than likely that the importance of the number of operations will have diminished considerably. Certainly, when aircraft noise levels fall below the general background level, there can be no impact from the number of operations, other than pure emotion about aircraft.

Figure 8.14 postulates that, although a correction of $10\log_{10} N$ is appropriate to areas affected by comparatively high levels of noise (i.e., those very close to the airport), a lower value of $5\text{–}7\log_{10} N$ will probably be more appropriate to the bulk of communities as noise levels fall progressively through to the end of this century.

Figure 8.14. The reducing importance of the number of operations as aircraft noise levels fall.

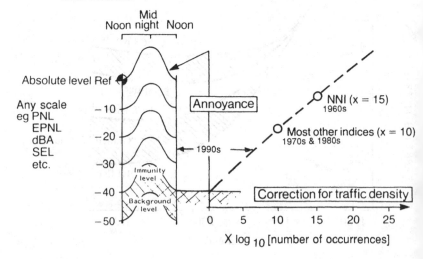

This being the case, then the likely overall trend in noise exposure (at constant noise technology) should be downwards for the next twenty years or so – perhaps not by as much as indicated by the lowest curve in Figure 8.13, but quite possibly by the amount indicated towards the top of the "graded" band. That would be a further improvement of about 60% of that already realised from the application of the turbofan.

This is an important conclusion, which, at first sight, suggests that the manufacturing industry can fold up its research programmes and sit back and wait for the application of high-bypass technology to solve the airport noise problem. Of course, this is extremely unlikely to reflect the true picture. New and expanding airports will always experience an increase in noise exposure. Moreover, what of new aircraft types and the new sound character that open rotors will generate? What about future supersonic aeroplanes? What about the wider use of helicopters? All these represent a substantial challenge in containing the situation, rather than improving it, for they lie outside the bounds of the generalised trend in noise versus bypass ratio given in Figure 8.10. For this reason, the industry worldwide has judged that it is not yet the time to cut back on noise research and, indeed, in several of the new technology areas, is expanding its activity.

Although we can hope that the situation will continue to improve, both the industry and government must plan ahead as far as a quarter of a century to ensure that the noise problem will not reemerge to the same degree as it did in the 1960s. This planning must take the form of structuring research programmes, efforts to control airport noise and international legislation to meet the needs of the twenty-first century. The developing fleet of new, quieter aircraft must be encouraged at the expense of the geriatrics, which must be phased out of service as quickly as possible.[51] The proposed moves in this direction in both Europe and the United States[30] must be enacted. The cost is by no means unacceptable to the business and, with the U.S. airlines[31] ready to cooperate in return for limited immunity at the airport, there must be some hope that positive action will result.

8.4 The direction of legislation

The foregoing sections have described the likely trends in airport noise through the next century. The achievement of an improved environment is as much a function of society's willingness to control the situation through legislation as it is the manufacturers' ability to make technological advances.

Noise legislation has grown out of hand, without any serious inspection of its continuing appropriateness to a changing airport noise climate. Since the original FAR Part 36 and ICAO Annex 16 were adopted in the early 1970s, there have been more amendments and modifications than new types of aircraft produced. In less than two decades ICAO has maintained active working groups that have been considering each class of aircraft and have held nine full-blown meetings on the subject and have scheduled a tenth for 1991. In the United States, FAR Part 36 has been modified on fifteen occasions and augmented by a technical "bible" of do's and don'ts in testing. With these changes, the political infant's insatiable appetite had caused it to grow from a simple "new aircraft-type" certification concept, affecting only the noisier jets, to consume all other jets, propeller aircraft (light and heavy separately), supersonic transports, and helicopters. In its time, and in its quest for power, it has also looked hungrily at the noise caused by thrust reversers and auxiliary power units. However, during the whole of this process of refinement, the one element that has never been addressed, with any positive outcome, is the relationship between noise certification and everyday operation, and the meaningfulness of the certification process in actually controlling noise at the airport. It is one thing to specify a series of standards the manufacturer has to meet, but quite another to determine a clear policy on the control of noise at the airport, and to act accordingly.

In the late 1970s, ICAO made an attempt to alter the basic framework of its noise certification scheme so that it would relate more closely to operational reality. However, all the various proposals were discounted on the basis that "the devil you know is better than the devil you don't". Despite this, and the fact that it was always stated publicly that noise certification was a tool for controlling source noise, not an operational (airport) tool, even the FAA fell into the trap of publishing listings of aircraft noise levels based on certificated data,[35] in a form that they thought might be useful to airport operators in determining their local operating rules. They even went so far as to devise night-operation strictures for their own Washington National Airport,[12] which were based on certificated noise numbers, albeit converted to the more generally applicable dBA scale.

As a result of all the attention it has received over the years, the noise certification process is now so complicated, time-consuming and expensive to the industry that one has to question its relevance. The three-point, fixed-distance, concept has outlived its usefulness – for one thing, some aircraft have become so quiet that measuring them at the original three reference points presents problems in interference from everyday background noise and, for another, the measuring distances are only representative of large airports. To be useful over the remainder of the century, the current noise certification structure needs to be dismantled, and only the successful technical components used to reassemble a system that is directed both at controlling airport noise and at providing an incentive to the manufacturer to do better than the standards demand. In my opinion, this would require an enlightened approach, which, unfortunately, is probably beyond the reach of established government thinking for it needs to embrace radical features, such as the abolition of the fixed-measurement distance and the noise-versus-number-of-engines-and-weight syndrome. The reasoning is as follows:

Although most nations have adopted the standards laid down in ICAO Annex 16 as a basic requirement to be met by the aircraft manufacturer, they all tend to devise their own separate control measures to deal with the problem at notoriously noise-sensitive airports. In some cases, national controls have been established. Examples include landing fees "graded" according to noise generated,[15] operational restrictions at night,[36] and refusals to permit older aircraft to be imported and added to the national register.[9] In establishing these subsidiary control measures, a popular indicator of noise impact is the size of a contour or constant noise level, reflecting the long-term noise exposure pattern at the airport, or the individual noise "footprint" that each aircraft movement stamps on the local community. Usually, there is a particular contour (set at a critical noise level) that defines the boundary between acceptability and unacceptability, and in some cases defines the way in which the public are treated. For example, in the United Kingdom, the 55 NNI contour is used to determine whether local residents can claim financial support in the cause of insulating their houses against aircraft noise[16] and the 35 NNI contour is generally regarded as the boundary of disturbance. Contour area was also used for ten years to define whether or not particular aircraft could use London's major airports at night and, recently, the concept has been extended for a further five years[36] to encourage wider use of the quieter types now entering service.

The use of contours to indicate a boundary of acceptability avoids some of the pitfalls of defining aircraft as "quiet" or "noisy" according to certification status, where absolute noise levels can vary widely in satisfying a standard whose critical dimensions are weight and number of engines. To say that aircraft that can comply with a particular certification requirement meet a superior technology standard and, by implication, those that cannot are unacceptably noisy causes no end of confusion. In the late 1970s, one U.S. airline satisfied its local airport requirements by replacing some older aircraft (those that could not meet the basic requirements of FAR 36) with aircraft that were compliant, and by implication environmentally acceptable, only to find that the new aircraft were, in fact, noisier in service and produced a higher complaint rate because they were physically bigger! It is easy to see how a fully laden 450-passenger quiet Boeing 747 at some 400 tonnes can cause far more annoyance than a noncompliant small but older aircraft with less than 100 seats. On take-off, the quiet Boeing 747 is allowed to generate up to 16 EPNdB more noise than the noisy smaller aircraft! This represents a threefold increase in annoyance and, clearly, is ridiculous when used as an operational bill of health. Equally odd is the fact that the same Boeing 747 can be deemed compliant with the latest requirements and operate from the same airports over the same routes as a DC10 that is up to 3 EPNdB quieter but is considered noncompliant because, in having one less engine, it has to satisfy a more stringent noise standard!

This is why noise certification and, in particular, its inappropriate application to the wider sphere of airport noise control, causes so many problems. One is forced to ask the fundamental question, "If contour-based methodologies are appropriate in addressing the real problem at the airport, why should they not be used in the noise certification process?" Arguments mustered in opposition to a contour-based process, apart from the fact that the noise business will have to get out of its established rut, are fairly tenuous. Some of them come from the industry, which has learned to live with FAR Part 36 and Annex 16 and the officials who administer them, and fear that change would automatically mean increased severity (as has usually been the case in the past). However, the majority of objections come from administrators, who argue that the public would not understand major change; but how much of the general public understands the complexities of noise certification in the first place? The only credible counterargument is that contours are extremely difficult to compute with accuracy, and even more difficult to validate experimentally (see Section 7.2). Hence, it follows that one could never hope to develop a certification scheme that relied on the precise definition of the detailed noise footprint from any given aircraft operation. To fully define an individual noise footprint, numerous measurements of the take-off and approach noise at different distances and different power settings would be necessary. This would be an extremely expensive business and would put even greater emphasis on the need for a stable test atmosphere than the current certification requirements do, with consequent increased delays to test programmes. Even so, this is no insuperable barrier to the use of the contour philosophy in certification, or indeed to the retention of the three basic measuring points, albeit on a "floating" basis.

8.5   The "pseudofootprint"

There are many variables to consider when computing the individual contour of constant annoyance, or footprint, associated with a single aircraft operation, but a standardised and idealised "pseudofootprint" could be calculated for the purposes of noise certification. The approach and take-off segments that comprise the total footprint of an aircraft operation into and out of an airport are shown in Figure 8.15, and they can be approximated simply and quite acceptably by ellipses. Three conditions have to be considered, take-off with full power, take-off with "cutback", and approach-to-landing. Cutback refers to operation at reduced engine power for noise abatement pur-

Figure 8.15. The noise footprint.

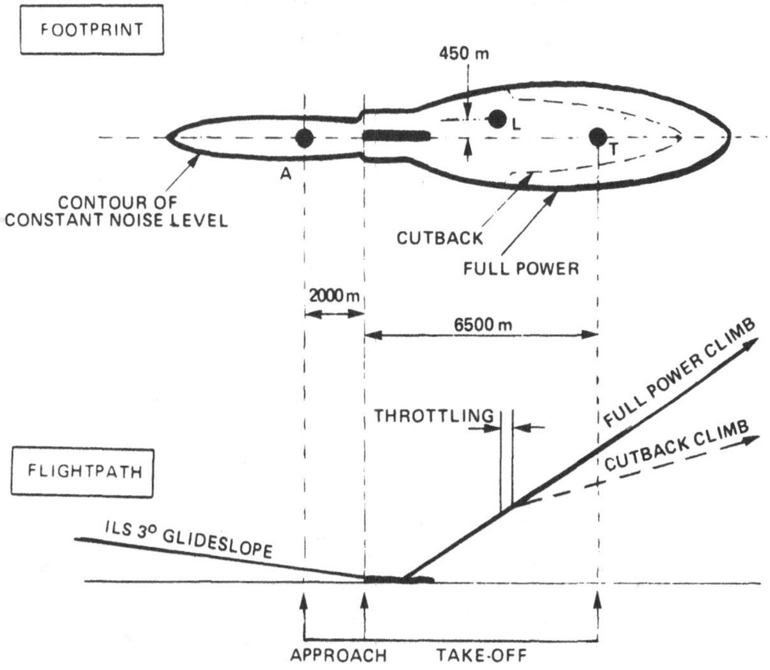

poses and is permitted in the legislation, once a safe altitude has been reached. As is shown in Figure 8.16, simple ellipses fit the shape of the (full-power) take-off and approach contours extremely well, but there is a divergence in the case of take-off cutback, where the slimming of the contour after power reduction is not well represented. On the other hand, the cutback procedure is a noise abatement device of somewhat dubious overall impact in the community. For one thing, high-bypass-ratio engines have a far "flatter" noise-versus-thrust relationship than older jets and, for another, any benefits from cutback are critically dependent upon local urban development patterns (see Section 2.4). Moreover, for air traffic control reasons, cutback is being used less and less, and, even where it is used, any lower noise levels in the vicinity of the airport as a result of power reduction are counterbalanced by a reduced aircraft climb path and, often, an increase in noise downstream of the airport when higher engine power is reestablished for the climb to cruising altitude. There is no magic in the process, for the total energy required to get the aircraft to cruise altitude remains the same no matter how the procedures are juggled.

Hence, any overestimation of the area made in a footprint-approximation exercise might well reflect the fact that the procedure is not worth all that it appears to be at first sight. On balance, therefore, the simple elliptical approximation of the pseudofootprint is probably more appropriate for the purpose of examining the relative annoyance

Figure 8.16. Approximation of footprint area with ellipses.

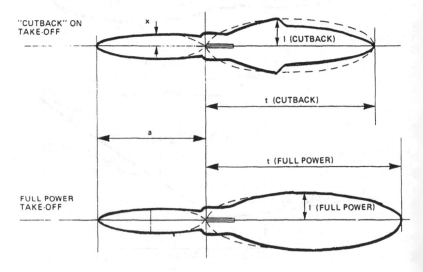

of different aircraft than the several current "indicators" derived from the certification process.

From Figure 8.16, it can be appreciated that all that would be necessary to establish the area of a pseudofootprint would be a determination of the "closure distances" to the longitudinal and lateral boundaries of the contour of a selected noise level, for both the take-off and approach cases. In practice, this would mean determining the length of both the take-off and approach ellipses (T and A), but only the width of the take-off contour (L). The approach contour, in being generated from the constant-speed, constant-thrust descent down a fixed 3° approach path, produces an ellipse of a regular aspect ratio, and the width is predetermined by the length and does not need measuring. The three closure distances that do need to be determined could be ascertained from tests via a simple modification to the current certification process – that is, by making the three measuring points "float" so as to determine the position at which any particular aircraft generated a chosen noise level, rather than by determining a noise level at a fixed distance, as is currently done. Providing the reference noise level was chosen sensibly, there would be few technical difficulties.

In fact, the benefits would extend to the technicalities of testing for compliance demonstration. For example, some of the laborious and detailed correction procedures and equivalencies that have been built up over the years could be abolished. The important tasks would be to anchor the policy in respect of the mode of certification control in a contour-based philosophy and to select the most appropriate noise level to act as the reference.

8.6   The "reference" noise level

Here, there are two interrelated factors to consider: the noise unit and the reference noise level. Any basic change in the certification scheme would allow authorities to consider changing the noise unit embodied in the process. Many people feel that the EPNdB has now outlived its usefulness and that a simpler and more widely appreciated unit is needed, preferably that recommended for noise-contour purposes – the sound exposure level, or SEL. This is a duration-corrected dBA-based measure, which, if used, would provide for better airport noise read-across since local airport noise restrictions and associated monitoring processes often adopt the dBA. But it is a separate debate and, for the moment, let us assume that the EPNL is to continue to be the definitive rating scale.

This being the case, the selection of the standard reference noise level of the scheme would have to be decided on the basis of the few important criteria:

relevance: In terms of the airport noise scene – by encompassing that proportion of the local population who can honestly be defined as affected adversely or "impacted" by aircraft noise.

sensibility: From the technical standpoint – high enough to ensure that there were no data loss or background noise problem in the measurement process, as is the case with current requirements.

meaningfulness: In terms of today's noise requirements – for continuity, that is, somewhere in the range of noise levels currently specified in the rules.

Since today's requirements span a wide range of EPNL values, between 89 and 108 (which represents a fourfold change in annoyance), this latter constraint is not particularly restrictive. However, from the point of view of ease of measurement, the value ought to be towards the upper end of the current scale rather than the lower end, to avoid the signal being corrupted by local ambient and data-system background noise. Consideration of a level that might be appropriate in terms of community reaction leads to the same conclusion; a figure as low as 90 EPNL only borders on the fringes of reality, in that it would produce large closure distances and extremely large and suspect contour areas; conversely, a level of 110 EPNdB is only experienced very close to the airport runway.

On balance, a value of 100 EPNL would seem to satisfy most criteria, since it would not produce noise contour areas so large as to be meaningless, or substantially different from any "real" contour that might be established from operational measurements.

8.7   Certification control

As to the mode of certification control, this could either be done by applying a constraint on the overall pseudofootprint area, or by the more direct application of upper limits to the distances from the airport at which 100 EPNdB is attained. The certification limit, either in terms of the closure-distance or footprint area, needs defining. That is the political problem.

If one looks at the numerical values that result from computing the pseudocontour area, or integrated noise area INA, at 100 EPNL for a range of aircraft, one arrives at the listing of the form shown in Table 8.3. Although only "guesstimates" from published noise data, there is a tremendous range of areas, from as high as several thousands of hectares, or tens of square kilometres, in the case of older aircraft, to less than 1 km^2 with modern types. This makes it clear just what the 19-dB spread of the current Annex 16 requirements means in terms of population affected by aircraft noise. To understand what such large area variations mean, it is necessary to plot these data in a different manner to that currently utilised, that is, not simply on an aircraft take-off weight scale, but in a way that reflects the performance capability of each aircraft out of a given airport and compares its noise with that of other aircraft using the same runways.

Table 8.3. *Construction of the integrated noise area metric*

Aircraft	Typical certification levels (EPNL)			Approx. 100 EPNL closures (km)			INA (km^2)
	Lateral	Take-off	Approach	l	t	a	
Concorde	116	110	116	2	10	11	50
Boeing 707	112	112	115	1.54	11	10	42
DC8-60	108	115	115	1	14	10	38
B747-100	103	110	114	0.77	10	9.5	27
Caravelle 10	102	97	107	0.54	6.0	6.0	10.7
B747-2/300	101	105	106	0.55	8.2	3.8	9.0
DC10-30	99	103	107	0.41	7.2	4.2	7.4
B727-200	103	100	103	0.58	6.5	2.8	7.1
DC9-50	103	97	102	0.58	4.7	2.5	5.3
DC10-40	96	100	105	0.32	6.5	3.4	5.0
B737-200	102	95	103	0.54	4.1	2.8	4.7
BAC1-11 3/400	103	94	102	0.60	3.9	2.5	4.7
L1011-100	95	97	103	0.25	5.0	2.5	3.0
F28-4000	98	101	101	0.38	2.9	2.3	2.6
B767	96	87	102	0.30	2.2	2.5	2.0
MD (DC9)-83	97	92	94	0.33	3.2	1.2	1.9
Airbus 300	92	90	102	0.20	2.6	2.3	1.6
B737-300	92	85	100	0.19	2.0	2.0	1.2
B757	93	85	95	0.22	2.0	1.2	0.9
Airbus 320	92	86	92	0.19	2.1	0.8	0.7
BAe 146	88	85	96	0.13	2.0	1.3	0.7
Fokker 100	90	86	94	0.16	2.1	1.1	0.7
Approximate ICAO Chapter 3 equivalents:							
Maximum limit	103	106	105	0.58	8.0	4.0	10
100-tonne trijet	98	96	102	0.38	5.0	2.3	4
Small twinjet	94	89	98	0.27	2.5	1.7	1.5

The simplest indicator of the size of airport that any given aircraft can utilise is the aircraft's "field length" under all-engines-operating conditions. This dictates the minimum distance in which an aircraft can take off or brake to a standstill after landing. In Figure 8.17, the 100 EPNdB INA has been plotted against aircraft field length. Naturally, those aircraft requiring the longest runways appear highest on the INA scale. These are usually the long-range three- and four-engined types, with a long take-off run. Here, some considerable contour area is attributable to the ground-roll régime and, because the climb performance of these aircraft is not sparkling, the take-off contour is extended even further. Also plotted on this illustration is a rough indication of where the original and latest certification requirements fall. The ragged representation of contour area afforded by the certification criteria results partly from the original decision to treat noise levels as a function of weight and partly from the 1977 modification to introduce the number of engines as a crude performance parameter. There is no clear relationship between airfield performance and the noise limits imposed. Had contour area been used as a basis for the original

Figure 8.17. Integrated noise areas compared with noise certification requirements.

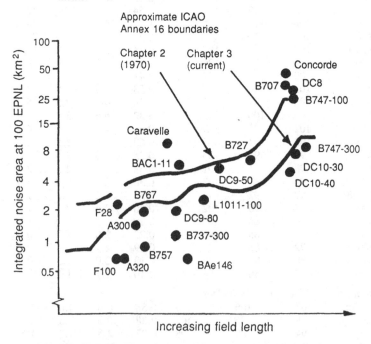

certification requirement, straight lines or second-order curves would have been drawn against field performance rather than aircraft weight, and some of the currently "compliant" aircraft would not have been awarded a noise certificate; equally, some "official" renegades might well be upright citizens. Such a situation might have better reflected general public opinion, for it is noise that awakens people and not the aircraft's pay-load, range, number of engines or certification category!

However, because of the wealth of experience gained and measured data taken, any new certification methodology would have to make allowances for decisions taken in the past. Therefore, in the first instance, if the new scheme were made applicable only to "new type designs", it could be objectively framed to stand the system in good stead for the future.

This could be accomplished by relating the maximum permitted closure distances to those achieved by Annex 16 (Chapter 3) compliant aircraft today, but on a field performance basis rather than on an aircraft take-off weight scale. Since field performance defines the smallest airport that an aircraft can use, it has a bearing on the proximity of the community in day-to-day operations and, hence, perceived annoyance. Short-runway aircraft would be expected to have smaller closure distances than aircraft operating solely from long runways at large international airports (see Fig. 8.18).

One useful advantage would be that there would no longer be any need for a trade-off allowance. This is today's escape clause for an aircraft that is critical under one of the three certification reference conditions but well-compliant under either of the others. Some aircraft have had to rely on the maximum trade allowance to meet the noise rules, with the result that they may be classified in a "quieter" bracket than they deserve. Under a performance-related scheme, if an aircraft failed to meet the standard, it would have to take a field-length penalty in operation, by an amount sufficient to allow it to meet the standard at the greater closure distance, whether it be sideline, take-off or approach. This could well exclude operations from critical (shorter) fields, and the pressure would certainly be on to "fix" the problem by quietening the aircraft. The clear advantage would be that no aircraft would fail to receive a noise certificate, but some might well be restricted to using only the largest airports, where they would not be offensive, until their noise had been reduced. Alternatively, like today, they could take a penalty via a reduced take-off weight, giving a better rate of climb and reducing the footprint closure distance.

8.8   Relationship to the airport

The areas plotted in Figure 8.17 actually do give some direct indication of overall annoyance, although they neglect the fact that there is an effectively "sterile" area owned by the airport, which is proportional in size to airport capacity, and which should be subtracted from the INA before the scale could be considered an absolute indicator of annoyance. Subtraction of the airport area from the equation would then give some indication of the number of people (or area of land) influenced by a particular aircraft in its everyday operation. As an example, in Figure 8.19, an area notionally attributable to the airport property limits has been plotted and the relative annoyance of different aircraft now becomes a function of how far any given aircraft INA protrudes beyond the airport property boundary. Interestingly, taken on an average basis, the aircraft INA values plotted fall substantially parallel to the property boundary line, indicating that perhaps a ratio of aircraft INA to notional airport area might be another way of

Figure 8.18. Certification standards set according to principal contour dimensions.

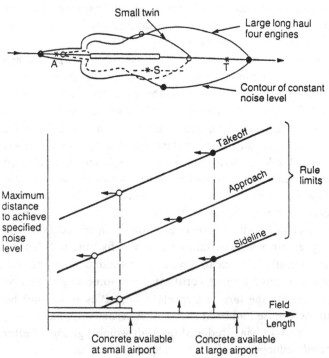

providing an indicator of annoyance. Perhaps this is a concept that might find favour with the airport operator?

8.9   The future – a summary

Common sense, or a "seat-of-the-pants feel" for the future airport noise situation, suggests that things are going to get progressively better as modern technology replaces the older engines in the airline fleet. Established noise exposure methodologies deny this by predicting an upsurge in the noise problem in about ten years' time. The reason for this is entirely the effect of the increasing number of aircraft in the world fleet. We shall have to wait to see whether the methodologies are wrong, or whether the aircraft noise control "business" is to flourish. In the meantime, the industry cannot afford to sit on its hands and travel hopefully, for it needs to ensure that its products satisfy future environmental demands. Noise is an operational shackle that needs to be cut away to allow the airlines to concentrate on more important issues.

If the problem does remain with us as the fleet grows, progressively

Figure 8.19. The relationship between integrated noise area and size of airport.

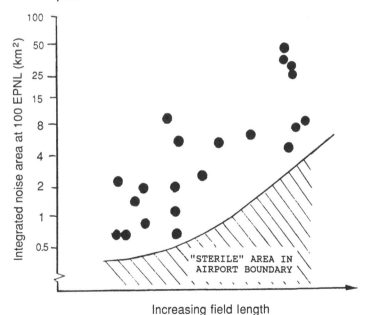

tougher, yet enlightened, legislation should be used to encourage competition and reward technical excellence in producing quieter aircraft. At present, it is left to those individual airport authorities who are at liberty to set their own noise limits to force that competition and reward the better product. Government should take a leaf from the local airports book and echo their more positive approach to the problem by updating legislation to meet the needs of the twenty-first century.

9

Review

The aircraft noise problem is an environmental "minus" that was born alongside the commercial gas turbine aeroengine more than thirty years ago. The issue came to a head in the 1960s, around the time when the number of jet-powered aircraft in the commercial fleet first exceeded the number of propeller-powered aircraft. During that decade, significant sums of money were expended in research and development exercises directed at quietening the exhaust noise of the subsonic fleet and in researching the noise of even higher-velocity jets that would have been a major problem if the supersonic transport had become a commercial success. Later in the 1960s, with the advent of the bypass-engine cycle, the emphasis was more on the noise generated inside the engine, with the large proportion of the extensive research funds necessary being provided by the taxpayers in those nations with aircraft- and engine-manufacturing capabilities.

Government action was not only limited to supplying the funds for research contracts, but major nations cooperated on a political front to develop noise certification requirements that were demanded of the manufacturers of all new aircraft produced from 1970 onwards. Coincidentally, technology moved forward and produced the high-bypass or turbofan-engine cycle, which, in reaping the benefits of the accelerated noise programmes of the 1960s, was only about one-quarter as noisy as the engines it replaced. Initially, however, the turbofan cycle was limited in its application to the new breed of larger wide-body jets, which had at least twice the passenger-carrying capacity of their forebears. As a result, the heavily utilised medium- and short-range sectors of the market remained powered by older engines with a high noise output. It was not until the 1980s, when, for sound commercial reasons (including meeting noise limits at airports), turbofan technology began to spread into this sector of the fleet and the benefits of the research work of the 1960s began to be felt fully, some twenty years on. Some of the older aircraft with many years of useful life ahead of them were, and still are being, reengined with turbofans.[44]

Even so, such is the working life of a commercial aeroplane that, even today as we look forward to the twenty-first century, still less than 50% of the fleet is made up of aircraft that are powered by turbofan engines. It will be well into the next century before all the older aircraft naturally work their way out of the fleet or are reengined with turbofans. By that time, and barring a world upheaval, the commercial fleet will have grown by over 50%, and this bigger fleet will be operating in ever-more congested flight patterns around the world's major airports. Where airports reach the limit of their capacity, new airport developments will bring the noise issue to a wider sector of the community.

Despite the reductions in absolute noise level that modern technology will bring to the high proportion of the fleet that is still "noisy", conventional techniques for predicting noise exposure do not foresee a continuing decline in the noise problem through to the next century. Indeed, they indicate an upswing in the problem in about ten years owing to the increased number of operations. It may well be, as has already been argued, that these predictions are not valid in the climate where individual aircraft noise levels are declining, but industry and government alike cannot afford to assume that the problem is now going to disappear. One major aircraft programme suffered badly from the assumption of success rather than having the technology available; that was the Concorde, an aircraft that would never have achieved noise certification under the original subsonic rules of 1970, let alone met today's regulations.

From industry's standpoint, the message has been well taken and there is ample evidence of continuing research programmes directed both at the conventional turbofan and new propulsion concepts, including the open rotor. From the government's standpoint, there must be a realisation that the controls hitherto exercised have been negatively structured and have failed to encourage competition throughout industry. They have not, however, failed to add to the industry's cost burden. The application of noise technology to the modern turbofan, at least sufficient to satisfy the latest certification standards, costs up to 1% in fuel consumption plus the loss of pay-load to the tune of one passenger and associated luggage per engine. In the mid-1970s, the largely cosmetic application of the original certification rules to the fleet of low-bypass-ratio-powered aircraft cost anything up to three times the fuel consumption penalty, by virtue of the aerodynamic losses caused by jet noise suppressors and internal nacelle treatment that had to be accommodated. It has been estimated that, combining the cost of the restrictions placed on operations, including the extra distance flown in using "quiet" corridors, with the costs incurred by the public and government, more than $1 billion changes hands each year in the pursuit of ways to minimise aircraft noise. Any new legislative initiatives from what has proved to be an administrative growth-industry in the past twenty years, must now be seen to be cost-effective. Those noise-control actions already taken that could be described as cost-effective, in that they have both reduced noise and encouraged competition, are restrictions imposed by local airport authorities, not generally by national governments. Here, the proof of quiet technology brings with it rewards in the form of greater freedom of operation.

The noise problem has thus far been limited to the developed and relatively prosperous nations. As air transport spreads its attractions to new areas, there will be increased awareness of the noise issue and environmental demands from nations hitherto passive on the subject. Whilst the industry bears a heavy responsibility for ensuring that its products are environmentally acceptable, both in existing and new markets, government bears the responsibility for ensuring that protec-

Figure 9.1. Age and distribution of the short- and medium-range "Stage 2" aircraft fleet.

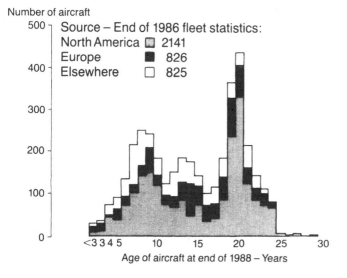

Number of aircraft

Source – End of 1986 fleet statistics:
North America 2141
Europe 826
Elsewhere 825

Age of aircraft at end of 1988 – Years

tion of the environment is encouraged directly, and that legal controls are exercised in an enlightened and rewarding manner. A suggested alternative approach to legislation has been given in Chapter 8.

In my opinion, whatever the predictive routines and the weighting factors embodied in the variety of existing exposure methodologies indicate about the future, the airport noise situation will continue to improve. Today, close on 4000 aircraft, or half of the total commercial fleet, consist of old and noisy medium-ranged aircraft, most of which are powered by the JT8D, the most numerous civil jet engine ever produced. However, as Figure 9.1 shows, around half of these have now used up a high proportion of their useful life and will be retired well before the end of the century. The remainder are currently a target for reengining with turbofans,[44] and one only has to recognise the way in which engines like the CFM-56, the IAE-V2500 and the RR-Tay will be replacing the JT8 in the coming years to realise that the problem of jet mixing noise will soon be a thing of the past. One day, jet noise may well be replaced by other sounds – low-frequency drones from open rotors or midfrequency whines from extremely high bypass-ratio turbofans – but the deafening roar of the jets of the 1960s should never be with us again. Legislation assures us of that; modern technology allows such an assurance to be made.

APPENDIX 1

Noise levels, scales and indices

The terms "level", "scale" and "index" are widely used in aircraft noise assessment. **Level** is taken to mean the quantity of noise, whether it be based purely on the physical measurement (e.g., sound pressure level) or subject to subsequent adjustment for human perception (e.g., perceived noise level). The level is measured against a **scale** (e.g., perceived noise level is assessed against the perceived noise scale, in units of PNdB). The term noise **index** refers to a numerical description of the impact of noise on the human being, taking into account not only the measured level and any adjustment for physical reaction, but also other factors such as duration and number of occurrences, and the ambient background level. The following are the most commonly used units, scales and indices:

deciBel (decibel): dB – The deciBel is a measure, on a logarithmic scale, of the magnitude of a particular sound intensity by reference to a standard quantity that represents the threshold of hearing. The sound pressure level L is, therefore, given by:

$$L = 10 \log_{10}(P/P_{ref})^2 \text{ dB}$$

or

$$20 \log_{10}(P/P_{ref}) \text{ dB},$$

where

$$P_{ref} = 20 \ \mu Pa.$$

Note: Table A.1.1 identifies the relationship between pressure, intensity and sound pressure level (SPL) for some common sounds.

Weighted sound pressure level – This is a sound pressure level weighted to reflect human interpretation of the loudness of sounds at different frequencies. Table A.1.2 identifies the A, B, C, D and N weightings.

A-Weighted SPL: dB(A) *or* L_A *or* dBA – This weighting relationship has been approximated on measuring instruments by a single electronic filtering network, and a close approximation of the A-weighted sound pressure level can be obtained directly by using many commercial sound-level meters.

B-Weighted SPL: dB(B) *or* L_B *or* dBB – The B weighting is similar to the A weighting but reflects human reaction to loudness at a level 40 dB higher. In practice it has little use, since it offers no clear advantage over the A weighting.

C-Weighted SPL: dB(C) *or* L_C *or* dBC – The C weighting is similar to both the A and B weightings except that it reflects human reaction to loudness at a level 30 dB higher than the B weighting (i.e., 70 dB higher than the A weighting). Again it offers no clear advantage over the A weighting network, but it has been used to describe sonic boom overpressures.

D-Weighted SPL: dB(D) *or* L_D *or* dBD – The D weighting exists solely for the purposes of describing aircraft noise and, unlike the A, B and C scales, it is based on annoyance rather than loudness and has been used in airport-monitoring systems as the basis of an approximation of PNdB.

N-Weighted SPL: dB(N) *or* dBN – The N weighting was established by the SAE in ARP 1080 in an attempt to produce an alternative to existing weightings of the late 1960s. It is based on the 40-Noy relationship

Table A.1.1. *Relationship between sound pressure, relative intensity and SPL*

Description of noise	Sound pressure (Pa)	Relative intensity	Sound pressure level, L (dB)
	200	10^{14}	140
Permanent hearing damage		10^{13}	130
Threshold of pain	20	10^{12}	120
Very noisy jet		10^{11}	110
Discothèque			
Steel riveter at 5 m	2	10^{10}	100
Inside noisy bus			
Noisy factory		10^9	90
Old subway train interior			
Noisy traffic in city	0.2	10^8	80
Loud party conversation			
Average factory		10^7	70
Supermarket			
Loud conversation	0.02	10^6	60
Busy office			
Average office		10^5	50
Living room			
Public park	0.002	10^4	40
Library			
Private office		10^3	30
Bedroom			
Whisper	0.0002	10^2	20
Soundproof room		10	10
Threshold of hearing (reference pressure)	0.00002	1	0

of the PNdB scale and follows closely the dBD weighting over its
(smaller) frequency range.

Noise level: NL – This is an A-weighted sound pressure level as measured
directly using a sound-level meter on "slow" response, as per ANSI
4-1983. It is specified by the California Department of Transport for
monitoring airport noise.

Perceived noise level: PNL *or* L_{PN} (in units of PNdB) – Again, this is a
weighted sound-pressure level, but in this case the weighting reflects
human annoyance to sounds at different frequencies and levels. The
scale was devised specifically for the purpose of describing aircraft
noise, and is the basis for the effective perceived noise level.

Table A.1.2. *Weighting schemes – based on loudness (A, B and C) and
annoyance (D and N)*

Frequency (Hz)	Correction (dB)				
	A	B	C	D	N
20	−50.5	−24.2	− 6.2	−20.6	—
25	−44.7	−20.4	− 4.4	−18.7	—
31.5	−39.4	−17.1	− 3.0	−16.7	—
40	−34.6	−14.2	− 2.0	−14.7	−14
50	−30.2	−11.6	− 1.3	−12.8	−12
63	−26.2	− 9.3	− 0.8	−10.9	−11
80	−22.5	− 7.4	− 0.5	− 9.0	− 9
100	−19.1	− 5.6	− 0.3	− 7.2	− 7
125	−16.1	− 4.2	− 0.2	− 5.5	− 6
160	−13.4	− 3.0	− 0.1	− 4.0	− 5
200	−10.9	− 2.0	0.0	− 2.6	− 3
250	− 8.6	− 1.3	0.0	− 1.6	− 2
315	− 6.6	− 0.8	0.0	− 0.8	− 1
400	− 4.8	− 0.5	0.0	− 0.4	0
500	− 3.2	− 0.3	0.0	− 0.3	0
630	− 1.9	− 0.1	0.0	− 0.5	0
800	− 0.8	0.0	0.0	− 0.6	0
1 000	0.0	0.0	0.0	0.0	0
1 250	0.6	0.0	0.0	2.0	2
1 600	1.0	0.0	− 0.1	4.9	6
2 000	1.2	− 0.1	− 0.2	7.9	8
2 500	1.3	− 0.2	− 0.2	10.4	10
3 150	1.2	− 0.4	− 0.5	11.6	11
4 000	1.0	− 0.7	− 0.8	11.1	11
5 000	0.5	− 1.2	− 1.3	9.6	10
6 300	− 0.1	− 1.9	− 2.0	7.6	9
8 000	− 1.1	− 2.9	− 3.0	5.5	6
10 000	− 2.5	− 4.3	− 4.4	3.4	3
12 500	− 4.3	− 6.1	− 6.2	1.4	0
16 000	− 6.6	− 8.4	− 8.5	− 0.7	—
20 000	− 9.3	−11.1	−11.2	− 2.7	—

Tone-corrected perceived noise level: PNLT *or* L_{TPN} (in units of PNdB) – This unit recognises the impact of a protrusive discrete tone on human annoyance by adding a penalty to the perceived noise level. It represents an intermediate step in computing EPNL.

Effective perceived noise level: EPNL *or* L_{EPN} (in units of EPNdB) – The EPNdB is the internationally recognised unit for describing the noise of a single aircraft operation. It was developed primarily for the purposes of aircraft noise certification and uses the earlier established perceived noise level as its basic noise measure. Additionally, it accounts for protrusive discrete tones via the PNLT and for the period during which the noise is 10 dB below its peak (tone-corrected) value, in what is referred to as a duration correction.

Note: The full calculation procedure for effective perceived noise level is given in Appendix 4.

Sound Exposure Level: SEL *or* L_{AE} – This is a dBA-based unit that accounts for duration in the same manner as the EPNdB but, unlike EPNdB, does not correct for discrete tone protrusion. It has been recommended for use in contour analyses by ICAO, ECAC and SAE.

Single-event noise exposure level: SENEL – A variant on the SEL, using a cut-off threshold level below which the energy is discounted. The main use is in noise monitoring, where a threshold trigger avoids the need for continuous recording.

Noise and number index: NNI – The NNI is of British origin and was probably the first noise exposure index to address the aircraft noise issue. It is based on the average (peak) perceived noise level over the daytime period 0700–1900, in the three summer months. It allows for the number of operations (N) by adding $15\log_{10} N$. It only includes events above 80 PNdB, a level that is considered to be the threshold of annoyance.

Isopsophic index: I – This French index is similar in concept to the NNI, but it covers a twenty-four-hour period with a night-time weighting of 6 or 10 dB, depending on the frequency of operations.

Day/night equivalent sound level: DNL *or* L_{DN} – This is an L_{eq}-type, dBA-based rating that is unique to the United States. The energy is averaged over a twenty-four-hour period but night-time events are weighted by the addition of 10 dBA.

Hourly noise level: HNL – Another U.S. (Californian) index, the HNL is based on the noise energy measured over the period of one hour.

Community noise equivalent level: CNEL *or* L_{CNE} – This is a variant on the DNL. It originated in California and was subsequently adopted by Denmark. Like the DNL, it makes use of the dBA, but with a night-time penalty of 10 dBA and also an evening weighting of 5 dBA.

Total noise load *or* **Kosten unit:** B – This is a rating developed in the Netherlands and is based on dBA. It subdivides the twenty-four-hour period into nine units, with a variety of weighting factors, depending upon sampled community response.

Noise exposure forecast: NEF – This was developed in the United States and was used by the federal authorities for many years in developing airport noise policies. It was based on the EPNL and drew a distinction between day and night-time operations.

Composite noise rating: CNR – The CNR is an NNI-type unit based on peak PNdB, but with a 13-dB night weighting.

Day/evening night level: DEN *or* L_{DEN} – This is a Danish unit, which, like the CNEL, is based on dBA but draws three distinctions – day, evening and night.

Equivalent level: $L_{eq}(A)$ – This is the unit recommended by the EEC, being a dBA-based L_{eq} derivative that can either be used on a twelve- or twenty-four-hour basis (with different normalisation corrections).

Storindex: Q – This is an L_{eq}-type unit that has a 5-dB weighting for night-time operation and finds favour in the Federal Republic of Germany. An Austrian version uses a 10-dB night weighting.

Weighted equivalent continuous perceived noise level: WECPNL *or* L_{WECPN} – This is the unit that was originally recommended by ICAO for inter-national harmonisation purposes. It is based on the EPNL and draws a distinction between day and night. In practice, it has found little favour other than in Japan, where it is used in a modified form, and in Italy and Brazil.

APPENDIX 2

Abbreviations and acronyms

a	speed of sound (sometimes)
AA	Administration de l'Aeronautique (Belgium)
AACC	Airport Associations Co-ordinating Council (see also AOCI)
AASC	Airworthiness Authorities Steering Committee (European)
ABNT	Associação Brasileira de Normas Técnicas
AEA	Association of European Airlines
AECMA	Association Européenne des Constructeurs de Matériel Aerospatial
AET	Airfields Environment Trust (United Kingdom)
AFNOR	Association Française de Normalisation
AGARD	Advisory Group for Aerospace Research and Development (European)
AI	Airbus Industrie
AIA	Aerospace Industries Association (variously applied in different countries; e.g., in Italy, Associazone Industrie Aerospaziali)
AIAA	American Institute of Aeronautics and Astronautics
AMD/BA	Avions Marcel Dassault-Breguet
ANCAT	Abatement of Nuisance Caused by Air Transport (ECAC noise subgroup)
ANOPP	Aircraft Noise Prediction Program (NASA Langley)
ANPRM	Advance Notice of Proposed Rule Making (United States)
ANSI	American National Standards Institute
ANTEWG	Aircraft Noise Technical Experts Working Group (EEC noise subgroup)
AOCI	Airport Operators Council International
ARA	Alfa Romeo Avio
AS	Société Nationale Industrielle Aerospatiale
ASME	American Society of Mechanical Engineers
ASTM	American Society for Testing and Materials
ATC	air traffic control
B	Kosten Unit – Dutch noise exposure index
BAe	British Aerospace plc
BCAC	Boeing Commercial Airplane Company

BCAR	British Civil Airworthiness Requirements – Section N is Noise
BDLI	Bundersverband der Deutschen Luft-und-Baumfar Industrie E.V
BIPM	International Bureau of Weights and Measures
BNA	Bureau of National Affairs (United States)
BPF	blade-passing frequency
BSI	British Standards Institute
c	speed of sound (other times)
CAA	Canadian Acoustical Association
CAA	Civil Aviation Authority/Administration (various)
CANAR	Consequences of Aircraft Noise Abatement Regulations – EEC contour programme
CAS	China Association for Standardisation
CEC	Commission of the European Communities (see EEC)
CEE	International Commission for Conformity Certification of Electrical Equipment (*also* IEC)
CNEL	Community Noise Equivalent Level (Californian index)
CNR	composite noise rating
cps	cycles per second (see also Hz)
dB	deciBel (commonly, decibel)
dB(A) or dBA	A-weighted sound level (also, B-, C-, D-weighted)
DCA	Directorate of Civil Aviation (Denmark)
DGAC	Direction Générale de l'Aviation Civile (France)
DGQ	Direcção-Geral da Qualidade (Portugal)
DIN	Deutsches Institut für Normung
DNL	day–night equivalent sound level – U.S. index
DoE	Department of the Environment (United Kingdom)
DOT	Department of Transportation (United States – parent body of FAA)
DS	Dansk Standardiseringsraad (Denmark)
DTI	Department of Trade and Industry (United Kingdom)
DTp	Department of Transport (United Kingdom)
ECAC	European Civil Aviation Conference (22 states)
EEC or EC	European (Economic) Community
EIA	Electronic Industries Association (United States)
ELOT	Hellenic Organisation for Standardisation
EPA	Environmental Protection Agency (United States)
EPNL	effective perceived noise level
FAA	Federal Aviation Administration (United States)
FAR Part 36	noise regulations of FARs
FARs	Federal Aviation Regulations (United States)
GARTEUR	European Group for Aeronautical Technological Research
GE	GE Aircraft Engines (General Electric subsidiary)
GIFAS	Groupement des Industries Françaises Aeronautiques et Spatiales
GOST	State Committee for Standards (Soviet Union)
HS	Hamilton Standard (Division of United Technologies)
HUD	Department of Housing and Urban Development (United States)

Hz	Hertz (frequency in cycles per second)
I	Isopsophic Index – French exposure index
IATA	International Air Transport Association
IBN	Institut Belge de Normalisation
ICAO	International Civil Aviation Organisation
ICCAIA	International Coordinating Council of Aerospace Industries Associations (AIAs of Europe, United States, Canada and Japan)
IEC	International Electrotechnical Commission
IEC	*See* CEE
IEEE	Institute of Electrical and Electronic Engineers (United States)
IIRS	Institute for Industrial Research and Standards (Eire)
INM	Integrated Noise Model – FAA contour programme and aircraft noise database
IoA	Institute of Acoustics (United Kingdom)
IRANOR	Instituto Español de Normalización
ISI	India Standards Institution
ISO	International Standards Organisation
ISVR	Institute of Sound and Vibration Research (of U.K. Southampton University)
JAEC	Japanese Aero Engine Company
JAR	Joint Airworthiness Requirements (European)
JISC	Japanese Industrial Standards Committee
KEBS	Kenya Bureau of Standards
kN	kilonewton – unit of force used in measuring thrust
L_{eq}	equivalent continuous sound level
L_x	. . . level, where x is the scale (e.g., L_{eq}, L_{PN}, L_{AX} etc.)
LAC	Lockheed Aircraft Corporation
LAX	A-weighted sound level (i.e., L_{AX}) *or* Los Angeles International Airport
LBA	Luftfahrt Bundesamt (Federal Republic of Germany)
MDC	McDonnell Douglas Corporation
NASA	National Aeronautics and Space Administration (United States)
NCF	Noise Control Foundation (United States)
NEF	noise exposure forecast – U.S. index
NNI	Nederlands Normalisatie-instituut
NNI	noise and number index – U.K. exposure index
NOISEMAP	U.S. Air Force noise contour program
NPL	National Physical Laboratory (United Kingdom)
NPRM	Notice of Proposed Rule Making (United States)
NSF	Norges Standardiseringforbund
OASPL	overall sound pressure level
OECD	Organisation for Economic Co-operation and Development (Europe)
ON	Osterreichisches Normungsinstitut
ONERA	Office Nationale d'Etudes et de Recherches Aeronautiques

Pa	pascal – unit of pressure, also called torr
PNL	perceived noise level
PNLT	tone-corrected PNL
PNYA	Port Authority of New York (and New Jersey)
PSI	Pakistan Standards Institute
PWA	Pratt and Whitney Aircraft (Division of United Technologies)
Q	Storindex – Federal German Republic noise exposure metric
RAI	Registrano Aeronautico Italiano
RLD	Rijksluchtvaartdienst (Netherlands Certificating Authority)
RR	Rolls-Royce plc
SAA	Standards Association of Australia
SABS	South Africa Bureau of Standards
SAE	Society of Automotive Engineers (United States)
SANZ	Standards Association of New Zealand
SCC	Standards Council of Canada
SEL	sound exposure level
SENEL	single-event noise exposure level
SFS	Suomen Standardisoimiliitt (Finland)
SII	Standards Institution of Israel
SIL	speech-interference level
SIRIM	Standards and Industrial Research of Malaysia
SIS	Standardiseringskommissionen i Sverige (Sweden)
SISIR	Singapore Institute of Standards and Industrial Research
SNECMA	Société Nationale d'Etude et de Construction de Moteurs d'Aviation
SPL	sound-pressure level
UNESCO	United Nations Educational, Scientific and Cultural Organisation
UT	United Technologies
WECPNL	weighted equivalent perceived noise level – ICAO's original recommended contour metric
WHO	World Health Organisation
YDNI	Badan Kerjasama Standardisasi (Indonesia)

APPENDIX 3

Useful addresses: partial listing of organisations concerned with noise in aerospace-manufacturing nations

Abatement of Nuisance Caused by Air Transport (ANCAT)
(Subgroup of ECAC – *see* ECAC)

Administration de l'Aeronautique (AA)
Direction Technique
Rue de la Fusée 90
B-1130 Brussels, Belgium

Advisory Group for Aerospace Research and Development (AGARD)
7 Rue Ancelle
F-92200, Neuilly-sur-Seine, France

Aerospace Industries Association (AIA)
1250 Eye Street, N.W.
Washington, D.C. 20005, USA

Airfields Environment Trust (AET)
17–19 Redcroft Way
London SE1 1TA, England

Airworthiness Authorities Steering Committee (AASC)
(Europe – *see* JAR)

Airbus Industrie (AI)
BP No. 33
31700 Blagnac, France

Airport Associations Coordinating Council (AACC)
P.O. Box 125, Ch-125
Geneva 15, Switzerland

American Institute for Aeronautics and Astronautics (AIAA)
1633 Broadway
New York, New York 10019, USA

American National Standards Institute (ANSI)
1430 Broadway
New York, New York 10018, USA

American Society of Mechanical Engineers (ASME)
345 East 47th Street
New York, New York 10017, USA

American Society for Testing and Materials (ASTM)
1916 Race Street
Philadelphia, Pennsylvania 19103, USA

Associaçao Brasileira de Normas Técnicas (ABNT)
Av. 13 de Maio, No. 13–28 andar
Caixa Postal 1680
CEP:20.003-Rio de Janeiro-RJ, Brazil

Association of European Airlines (AEA)
Avenue Lousie 350, Bt 4
B-1050 Brussels, Belgium

Association Européenne des Constructeurs de Matériel Aerospatial (AECMA)
88 Boulevarde Malesherbes
F-75008, Paris, France

Association Française de Normalisation (AFNOR)
Tour Europe, Cedex 7
92080 Paris La Defense, France

Association Suisse de Normalisation (SNV)
Kirchenweg 4, Postfa
8032 Zurich, Switzerland

Associazione Industrie Aerospaziali (AIA)
Via Nazionale 200
00184 – Rome, Italy

Avions Marcel Dassault-Breguet Aviation (AMD/BA)
Direction Technique, BP 24
33701 Merignac, France

Badan Kerjasama Standardisasi LIPI-YDNI (YDNI)
Jln. Teuku Chik Ditiro 43
P.O. Box 250
Jakarta, Indonesia

Boeing Commercial Airplane Company (BCAC)
P.O. Box 3707
Seattle, Washington 98124, USA

British Aerospace plc (BAe)
Civil Aircraft Division
Barnet Bypass, Hatfield
Hertfordshire, AL10 9TL, England

British Standards Institute (BSI)
2 Park Street
London W1A 2BS, England

Bureau International des Poid et Mesures (BIPM)
Pavillon de Breteuil
92310 Sèvres, France

Bureau of National Affairs, Inc. (BNA)
(Publishers of *Noise Regulation Reporter*)
1231 25th Street, N.W.
Washington, D.C. 20037, USA

Canadian Acoustical Association (CAA)
P.O. Box 3651, Station C
Ottawa, Ontario K1Y 4J7, Canada

China Association for Standardisation (CAS)
P.O. Box 820
Beijing, China

Civil Aviation Authority (CAA)
Aviation House
South Area, Gatwick Airport
West Sussex RH6 OYR, England

Commission of the European Communities (CEC, EEC or EC)
Rue de la Loi
B-1049 Brussels, Belgium

Dansk Standardiseringsraad (DS)
Aurehojvej 12, Postbox 77
DK-2900 Hellerup, Denmark

Department of the Environment (DOE)
2 Marsham Street
London SW1P 3EB, England

Department of Transport (DTp)
2 Marsham Street
London SW1P 3EB, England

Department of Transportation (DOT)
Parent Department of FAA – *see*
FAA

Deutsches Institut für Normung (DIN)
Burggrafenstrasse 4–10, Postfach
1107
D-1000 Berlin 30, Federal Republic
of Germany

Direcção-Geral da Qualidade (DGQ)
Rua Jose Estevao, 83-A
1199 Lisboa Codex, Portugal

*Direction Generale de l'Aviation
Civile (DGAC)*
246 Rue Lecourbe
75732 Paris Cedex 15, France

Directorate of Civil Aviation (DCA)
P.O. Box 744
DK-2450, Copenhagen, Denmark

*Electronic Industries Association
(EIA)*
2001 Eye Street, N.W.
Washington D.C. 20006, USA

*Ente Nazionale Italiano di
Unificazione (UNI)*
Piazza Armando Diaz 2
1-20123 Milano, Italy

*Environmental Protection Agency
(EPA)*
Office of Environmental Control
401 Main Street, S.W.
Washington, D.C. 20460, USA

*European Civil Aviation Conference
(ECAC)*
3 Bis Villas Emile Bergerat
F-92522, Neuilly-sur-Seine, France

*Federal Aviation Administration
(FAA)*
800 Independence Ave., S.W.
Washington, D.C. 20591, USA

Federal Ministry of Transport (LBA)
Department of Civil Aviation
Kennedyallee, Postfach 20 0100
D-5300 Bonn 2, Federal Republic of
Germany

Fokker Aircraft-BV (Fokker)
Postbus 7600
1117 ZJ, Schiphol, The Netherlands

GE Aircraft Engines (GE)
One Neumann Way
Cincinnati, Ohio 45220, USA

*Groupement des Industries Françaises
Aeronautiques et Spatiales (GIFAS)*
4 Rue Galilee
75782 Paris Cedex 16, France

*Hamilton Standard
(HS, Division of UTC)*
Bradley Field Road
Windsor Locks, Connecticut 06096,
USA

*Hellenic Organization for
Standardization (ELOT)*
Didotou 15
106 80 Athens, Greece

India Standards Institution (ISI)
Manak Bhavan
9 Bejadur Shah Zafar Marg
New Delhi 110002, India

Institut Belge de Normalisation (IBN)
Av. de la Brabanconne, 29
B-1040, Bruxelles, Belgium

Institute of Acoustics (IoA)
25 Chambers Street
Edinburgh EH1 1HU, Scotland

*Institute of Electrical and Electronic
Engineers (IEEE)*
345 East 47th Street
New York, New York 10017, USA

*Institute for Industrial
Research and Standards (IIRS)*
Ballymum Road
Dublin – 9, Eire

*Institute of Sound and Vibration
Research (ISVR)*
University of Southampton
Highfield
Southampton SO9 5NH, England

*Instituto Español de Normalización
(IRANOR)*
Calle Fernandex de la Hoz, 52
Madrid 10, Spain

*Institutul Roman de Standardizare
(IRS)*
Casuta Postala 63–87
Bucharest 1, Rumania

*International Air Transport
Association (IATA)*
2000 Peel Street
Montréal, Québec H3A 2R4, Canada

*International Civil Aviation
Organisation (ICAO)*
1000 Sherbrooke Street West, Suite
400
Montréal, Québec H3A 2R2, Canada

*International Commission for
Conformity Certification of Electrical Equipment (Commission
Internationale de Certification
de Conformité de l'Equipement
Electrique) (IEC, or CEC)*
Utrechtsecrey 310
6812 AR Arnhem, The Netherlands

*International Co-ordinating Council
of Aerospace Industries
Associations (ICCAIA)*
1250 Eye Street, N.W.
Washington, D.C. 20005, USA

*International Electrotechnical
Commission (IEC)*
3 Rue de Varembe
1211 Genève 20, Switzerland

*International Standards Organisation
(ISO)*
1 Rue de Varembe
1211 Genève 20, Switzerland

*Japanese Industrial Standards
Committee (JISC)*
c/o Standards Department
Agency of Industrial Science and
Technology
Ministry of International Trade and
Industry
1-3-1, Ksumigaseki
Chiyoda-ku, Tokyo 100, Japan

*Joint Airworthiness Requirements
Secretariat (JARS)*
Aviation House
South Area, Gatwick Airport
West Sussex RH6 OYR, England

Kenya Bureau of Standards (KEBS)
Off Mombasa Road
Behind Belle Vue Cinema
P.O. Box 54974
Nairobi, Kenya

*Lockheed Aircraft Corporation
(LAC)*
P.O. Box 551
Burbank, California 91503, USA

*McDonnell Douglas Corporation
(MDC)*
3855 Lakewood Boulevard
Long Beach, California 90846, USA

*National Aeronautics and Space
Administration (NASA)*
Headquarters
600 Independence Ave, S.W.
Washington, D.C. 20546, USA
Regional Research Establishments:
Ames Research Center
MS 247–1
Moffett Field, California 99035, USA
Lewis Research Center
Cleveland, Ohio 44135, USA
Langley Research Center
MS 461
Hampton, Virginia 23665, USA

National Physical Laboratory (NPL)
Queens Road, Teddington
Middlesex TW11 0LW, England

Nederlands Normalisatie-institut (NNI)
Kalfjeslaan 2
P.O. Box 5059
2600 GB Delft, The Netherlands

Noise Control Foundation (NCF)
P.O. Box 2469
Arlington Branch
Poughkeepsie, New York 12603, USA

Norges Standardiseringsforbund (NSF)
Postboks 7020 Homansbyen
N-Oslo 3, Norway

Office National d'Etudes et de Recherches Aeronautiques (ONERA)
29 Avenue de la Division Leclerc
92-Chatillon, Paris, France

Organisation for Economic Co-operation and Development (OECD)
2 Rue André Pascal
Paris 16, France

Osterreichisches Normungsinstitut (ON)
Heinestrasse 38
Postfact 130
A-1021 Wien, Austria

Pakistan Standards Institution (PSI)
39 Garden Road
Sadda, Karachi-3, Pakistan

Pratt and Whitney Aircraft (PWA)
400 Main Street
East Hartford, Connecticut 06108, USA

Product Standards Agency (PSA)
Ministry of Trade and Industry
361 Sen. Gil J. Puyat Avenue
Makat, Metro Manila 3117
Manila, Philippines

Registro Aeronautico Italiano (RAI)
Corso di Porto Romana 46
20122 Milano, Italy

Rijksluchtvaartinspectie (RLD)
Postbus 7555
1117 ZH Schiphol, The Netherlands

Rolls-Royce plc (RR)
65 Buckingham Gate
London SW1 6AT, England

Short Brothers plc (Shorts)
P.O. Box 241
Airport Road
Belfast BT3 9DZ, Northern Ireland

Singapore Institute of Standards and Industrial Research (SISIR)
Maxwell Road
P.O. Box 2611
Singapore 9046

Société Nationale Industrielle Aerospatiale (AS)
37 Boulevard de Montmorency
75 Paris, France

Society of Automotive Engineers (SAE)
400 Commonwealth Drive
Warrendale, Pennsylvania 15096, USA

South African Bureau of Standards (SABS)
Private Bag X191
Pretoria 0001, Republic of South Africa

Standardiseringskommissionen i Sverige (SIS)
Tegnergatan 11
Box 3 295
S-103 66 Stockholm, Sweden

Standards Association of Australia (SAA)
Standards House
80–86 Arthur Street
North Sydney, NSW 2060, Australia

Standards Association of New Zealand (SANZ)
Private Bag
Wellington, New Zealand

Standards Council of Canada (SCC)
International Standardization Branch
2000 Argentia Road, Suite 2-401
Mississauga, Ontario, Canada

Standards and Industrial Research
Institute of Malaysia (SIRIM)
Lot 10810, Phase 3
Federal Highway, P.O. Box 35
Shah Alam, Selangor, Malaysia

Standards Institution of Israel (SII)
42 University Street
Tel Aviv 69977, Israel

Suomen Standardisoimiliitt r.y. (SFS)
P.O. Box 205
SF-0012 Helsinki 12, Finland

United Nations Educational, Scientific
and Cultural Organisation
(UNESCO)
7 Place de Fontenoy
F-75700, Paris, France

USSR State Committee for Standards
(GOST)
Leninsky Prospekt 9
Moskva 117049, USSR

World Health Organisation (WHO)
Avenue Appia
1211 Genève, Switzerland

APPENDIX 4

Calculation of effective perceived noise-level data from measured noise data: excerpts from ICAO Annex 16*

4.1 General

4.1.1 The basic element in the noise certification criteria shall be the noise evaluation measure designated effective perceived noise level, EPNL, in units of EPNdB, which is a single number evaluator of the subjective effects of aeroplane noise on human beings. Simply stated, EPNL shall consist of instantaneous perceived noise level, PNL, corrected for spectral irregularities (the correction, called "tone correction factor", is made for the maximum tone only at each increment of time) and for duration.

4.1.2 Three basic physical properties of sound pressure shall be measured: level, frequency distribution, and time variation. More specifically, the instantaneous sound pressure level in each of 24 one-third octave bands of the noise shall be required for each 500 ms increment of time during the aeroplane flyover.

4.1.3 The calculation procedure which utilizes physical measurements of noise to derive the EPNL evaluation measure of subjective response shall consist of the following five steps:

a) the 24 one-third octave bands of sound pressure level are converted to perceived noisiness by means of a noy table. The noy values are combined and then converted to instantaneous perceived noise levels, PNL(k);

b) a tone correction factor, C(k) is calculated for each spectrum to account for the subjective response to the presence of spectral irregularities;

* Reprinted by permission.

Table 2-1. Perceived noisiness (noys) as a function of sound pressure level

One-third Octave band centre frequencies (Hz)

SPL	50	63	80	100	125	160	200	250	315	400	500	630	800	1000	1250	1600	2000	2500	3150	4000	5000	6300	8000	10000
4																			0.10					
5																		0.10	0.11	0.10				
6																		0.11	0.12	0.11	0.10			
7																		0.12	0.14	0.13	0.11			
8																		0.14	0.16	0.14	0.13			
9																	0.10	0.16	0.17	0.16	0.14			
10																	0.11	0.17	0.19	0.18	0.16	0.10		
11																	0.13	0.19	0.22	0.21	0.18	0.12		
12																0.10	0.14	0.22	0.24	0.24	0.21	0.14		
13																0.11	0.16	0.24	0.27	0.27	0.24	0.16		
14																0.13	0.18	0.27	0.30	0.30	0.27	0.19		
15															0.10	0.14	0.21	0.30	0.33	0.33	0.30	0.22		
16										0.10	0.10	0.10	0.10	0.10	0.11	0.16	0.24	0.33	0.35	0.35	0.33	0.26		
17										0.11	0.11	0.11	0.11	0.11	0.13	0.18	0.27	0.35	0.38	0.38	0.35	0.30	0.10	
18									0.10	0.13	0.13	0.13	0.13	0.13	0.15	0.21	0.30	0.38	0.41	0.41	0.38	0.33	0.12	
19									0.11	0.14	0.14	0.14	0.14	0.14	0.17	0.24	0.33	0.41	0.45	0.45	0.41	0.36	0.14	
20									0.13	0.16	0.16	0.16	0.16	0.16	0.20	0.27	0.36	0.45	0.49	0.49	0.45	0.39	0.17	
21								0.10	0.14	0.18	0.18	0.18	0.18	0.18	0.23	0.30	0.39	0.49	0.53	0.53	0.49	0.42	0.21	0.10
22								0.11	0.16	0.21	0.21	0.21	0.21	0.21	0.26	0.33	0.42	0.53	0.57	0.57	0.53	0.46	0.25	0.11
23								0.13	0.18	0.24	0.24	0.24	0.24	0.24	0.30	0.36	0.46	0.57	0.62	0.62	0.57	0.50	0.30	0.13
24							0.10	0.14	0.21	0.27	0.27	0.27	0.27	0.27	0.33	0.40	0.50	0.62	0.67	0.67	0.62	0.55	0.33	0.15
25							0.11	0.16	0.24	0.30	0.30	0.30	0.30	0.30	0.35	0.43	0.55	0.67	0.73	0.73	0.67	0.60	0.36	0.17
26							0.13	0.18	0.27	0.33	0.33	0.33	0.33	0.33	0.38	0.48	0.60	0.73	0.79	0.79	0.73	0.65	0.39	0.20
27						0.10	0.14	0.21	0.30	0.35	0.35	0.35	0.35	0.35	0.41	0.52	0.65	0.79	0.85	0.85	0.79	0.71	0.42	0.23
28						0.11	0.16	0.24	0.33	0.38	0.38	0.38	0.38	0.38	0.45	0.57	0.71	0.85	0.92	0.92	0.85	0.77	0.46	0.26
29						0.13	0.18	0.27	0.35	0.41	0.41	0.41	0.41	0.41	0.49	0.63	0.77	0.92	1.00	1.00	0.92	0.84	0.50	0.30

302

30					0.10	0.14	0.21	0.30	0.38	0.45	0.45	0.45	0.45	0.45	0.45	0.53	0.69	0.84	1.00	1.07	1.07	1.00	0.92	0.55	0.33
31					0.11	0.16	0.24	0.33	0.41	0.49	0.49	0.49	0.49	0.49	0.49	0.57	0.76	0.93	1.07	1.15	1.15	1.07	1.00	0.60	0.37
32					0.13	0.18	0.27	0.36	0.45	0.53	0.53	0.53	0.53	0.53	0.53	0.62	0.83	1.00	1.15	1.23	1.23	1.15	1.07	0.65	0.41
33					0.14	0.21	0.30	0.39	0.49	0.57	0.57	0.57	0.57	0.57	0.57	0.67	0.91	1.07	1.23	1.32	1.32	1.23	1.15	0.71	0.45
34					0.16	0.24	0.33	0.42	0.53	0.62	0.62	0.62	0.62	0.62	0.62	0.73	1.00	1.15	1.32	1.41	1.41	1.32	1.23	0.77	0.50
35				0.11	0.18	0.27	0.36	0.46	0.57	0.67	0.67	0.67	0.67	0.67	0.67	0.79	1.07	1.23	1.41	1.51	1.51	1.41	1.32	0.84	0.55
36				0.13	0.21	0.30	0.40	0.50	0.62	0.73	0.73	0.73	0.73	0.73	0.73	0.85	1.15	1.32	1.51	1.62	1.62	1.51	1.41	0.92	0.61
37				0.15	0.24	0.33	0.43	0.55	0.67	0.79	0.79	0.79	0.79	0.79	0.79	0.92	1.23	1.41	1.62	1.74	1.74	1.62	1.51	1.00	0.67
38				0.17	0.27	0.37	0.48	0.60	0.73	0.85	0.85	0.85	0.85	0.85	0.85	1.00	1.32	1.51	1.74	1.86	1.86	1.74	1.62	1.10	0.74
39				0.20	0.30	0.41	0.52	0.65	0.79	0.92	0.92	0.92	0.92	0.92	0.92	1.07	1.41	1.62	1.86	1.99	1.99	1.86	1.74	1.21	0.82
40			0.12	0.23	0.33	0.45	0.57	0.71	0.85	1.00	1.00	1.00	1.00	1.00	1.00	1.15	1.51	1.74	1.99	2.14	2.14	1.99	1.86	1.34	0.90
41			0.14	0.26	0.37	0.50	0.63	0.77	0.92	1.07	1.07	1.07	1.07	1.07	1.07	1.23	1.62	1.86	2.14	2.29	2.29	2.14	1.99	1.48	1.00
42			0.16	0.30	0.41	0.55	0.69	0.84	1.00	1.15	1.15	1.15	1.15	1.15	1.15	1.32	1.74	1.99	2.29	2.45	2.45	2.29	2.14	1.63	1.10
43			0.19	0.33	0.45	0.61	0.76	0.92	1.07	1.23	1.23	1.23	1.23	1.23	1.23	1.41	1.86	2.14	2.45	2.63	2.63	2.45	2.29	1.79	1.21
44			0.22	0.37	0.50	0.67	0.83	1.00	1.15	1.32	1.32	1.32	1.32	1.32	1.32	1.52	1.99	2.29	2.63	2.81	2.81	2.63	2.45	1.99	1.34
45		0.12	0.26	0.42	0.55	0.74	0.91	1.08	1.24	1.41	1.41	1.41	1.41	1.41	1.41	1.62	2.14	2.45	2.81	3.02	3.02	2.81	2.63	2.14	1.48
46		0.14	0.30	0.46	0.61	0.82	1.00	1.16	1.33	1.52	1.52	1.52	1.52	1.52	1.52	1.74	2.29	2.63	3.02	3.23	3.23	3.02	2.81	2.29	1.63
47		0.16	0.34	0.52	0.67	0.90	1.08	1.25	1.42	1.62	1.62	1.62	1.62	1.62	1.62	1.87	2.45	2.81	3.23	3.46	3.46	3.23	3.02	2.45	1.79
48		0.19	0.38	0.58	0.74	1.00	1.17	1.34	1.53	1.74	1.74	1.74	1.74	1.74	1.74	2.00	2.63	3.02	3.46	3.71	3.71	3.46	3.23	2.63	1.98
49		0.22	0.43	0.65	0.82	1.08	1.26	1.45	1.64	1.87	1.87	1.87	1.87	1.87	1.87	2.14	2.81	3.23	3.71	3.97	3.97	3.71	3.46	2.81	2.18
50	0.12	0.26	0.49	0.72	0.90	1.17	1.36	1.56	1.76	2.00	2.00	2.00	2.00	2.00	2.00	2.30	3.02	3.46	3.97	4.26	4.26	3.97	3.71	3.02	2.40
51	0.14	0.30	0.55	0.80	1.00	1.26	1.47	1.68	1.89	2.14	2.14	2.14	2.14	2.14	2.14	2.46	3.23	3.71	4.26	4.56	4.56	4.26	3.97	3.23	2.63
52	0.17	0.34	0.62	0.90	1.08	1.36	1.58	1.80	2.03	2.30	2.30	2.30	2.30	2.30	2.30	2.64	3.46	3.97	4.56	4.89	4.89	4.56	4.26	3.46	2.81
53	0.21	0.39	0.70	1.00	1.18	1.47	1.71	1.94	2.17	2.46	2.46	2.46	2.46	2.46	2.46	2.83	3.71	4.26	4.89	5.24	5.24	4.89	4.56	3.71	3.02
54	0.25	0.45	0.79	1.09	1.28	1.58	1.85	2.09	2.33	2.64	2.64	2.64	2.64	2.64	2.64	3.03	3.97	4.56	5.24	5.61	5.61	5.24	4.89	3.97	3.23
55	0.30	0.51	0.89	1.15	1.35	1.71	2.00	2.25	2.50	2.83	2.83	2.83	2.83	2.83	2.83	3.25	4.26	4.89	5.61	6.01	6.01	5.61	5.24	4.26	3.46
56	0.34	0.59	1.00	1.29	1.50	1.85	2.15	2.42	2.69	3.03	3.03	3.03	3.03	3.03	3.03	3.48	4.56	5.24	6.01	6.44	6.44	6.01	5.61	4.56	3.71
57	0.39	0.67	1.09	1.40	1.63	2.00	2.33	2.61	2.88	3.25	3.25	3.25	3.25	3.25	3.25	3.73	4.89	5.61	6.44	6.90	6.90	6.44	6.01	4.89	3.97
58	0.45	0.77	1.18	1.53	1.77	2.15	2.51	2.81	3.10	3.48	3.48	3.48	3.48	3.48	3.48	4.00	5.24	6.01	6.90	7.39	7.39	6.90	6.44	5.24	4.26
59	0.51	0.87	1.29	1.66	1.92	2.33	2.71	3.03	3.32	3.73	3.73	3.73	3.73	3.73	3.73	4.29	5.61	6.44	7.39	7.92	7.92	7.39	6.90	5.61	4.56

Table 2-1. (*Cont.*)

SPL	50	63	80	100	125	160	200	250	315	400	500	630	800	1000	1250	1600	2000	2500	3150	4000	5000	6300	8000	10000
60	0.59	1.00	1.40	1.81	2.08	2.51	2.93	3.26	3.57	4.00	4.00	4.00	4.00	4.00	4.59	6.01	6.90	7.92	8.49	8.49	7.92	7.39	6.01	4.89
61	0.67	1.10	1.53	1.97	2.26	2.71	3.16	3.51	3.83	4.29	4.29	4.29	4.29	4.29	4.92	6.44	7.39	8.49	9.09	9.09	8.49	7.92	6.44	5.24
62	0.77	1.21	1.66	2.15	2.45	2.93	3.41	3.78	4.11	4.59	4.59	4.59	4.59	4.59	5.28	6.90	7.92	9.09	9.74	9.74	9.09	8.49	6.90	5.61
63	0.87	1.32	1.81	2.34	2.65	3.16	3.69	4.06	4.41	4.92	4.92	4.92	4.92	4.92	5.66	7.39	8.49	9.74	10.4	10.4	9.74	9.09	7.39	6.01
64	1.00	1.45	1.97	2.54	2.88	3.41	3.98	4.38	4.73	5.28	5.28	5.28	5.28	5.28	6.06	7.92	9.09	10.4	11.2	11.2	10.4	9.74	7.92	6.44
65	1.11	1.60	2.15	2.77	3.12	3.69	4.30	4.71	5.08	5.66	5.66	5.66	5.66	5.66	6.50	8.49	9.74	11.2	12.0	12.0	11.2	10.4	8.49	6.90
66	1.22	1.75	2.34	3.01	3.39	3.99	4.64	5.07	5.45	6.06	6.06	6.06	6.06	6.06	6.96	9.09	10.4	12.0	12.8	12.8	12.0	11.2	9.09	7.39
67	1.35	1.92	2.54	3.28	3.68	4.30	5.01	5.46	5.85	6.50	6.50	6.50	6.50	6.50	7.46	9.74	11.2	12.8	13.8	13.8	12.8	12.8	9.74	7.92
68	1.49	2.11	2.77	3.57	3.99	4.64	5.41	5.88	6.27	6.96	6.96	6.96	6.96	6.96	8.00	10.4	12.0	13.8	14.7	14.7	13.8	12.8	10.4	8.49
69	1.65	2.32	3.01	3.88	4.33	5.01	5.84	6.33	6.73	7.46	7.46	7.46	7.46	7.46	8.57	11.2	12.8	14.7	15.8	15.8	14.7	13.8	11.2	9.09
70	1.82	2.55	3.28	4.23	4.69	5.41	6.31	6.81	7.23	8.00	8.00	8.00	8.00	8.00	9.19	12.0	13.8	15.8	16.9	16.9	15.8	14.7	12.0	9.74
71	2.02	2.79	3.57	4.60	5.09	5.84	6.81	7.33	7.75	8.57	8.57	8.57	8.57	8.57	9.85	12.8	14.7	16.9	18.1	18.1	16.9	15.8	12.8	10.4
72	2.23	3.07	3.88	5.01	5.52	6.31	7.36	7.90	8.32	9.19	9.19	9.19	9.19	9.19	10.6	13.8	15.8	18.1	19.4	19.4	18.1	16.9	13.8	11.2
73	2.46	3.37	4.23	5.45	5.99	6.81	7.94	8.50	8.93	9.85	9.85	9.85	9.85	9.85	11.3	14.7	16.9	19.4	20.8	20.8	19.4	18.1	14.7	12.0
74	2.72	3.70	4.60	5.94	6.50	7.36	8.57	9.15	9.59	10.6	10.6	10.6	10.6	10.6	12.1	15.8	18.1	20.8	22.3	22.3	20.8	19.4	15.8	12.8
75	3.01	4.06	5.01	6.46	7.05	7.94	9.19	9.85	10.3	11.3	11.3	11.3	11.3	11.3	13.0	16.9	19.4	22.3	23.9	23.9	22.3	20.8	16.9	13.8
76	3.32	4.46	5.45	7.03	7.65	8.57	9.85	10.6	11.0	12.1	12.1	12.1	12.1	12.1	13.9	18.1	20.8	23.9	25.6	25.6	23.9	22.3	18.1	14.7
77	3.67	4.89	5.94	7.66	8.29	9.19	10.6	11.3	11.8	13.0	13.0	13.0	13.0	13.0	14.9	19.4	22.3	25.6	27.4	27.4	25.6	23.9	19.4	15.8
78	4.06	5.37	6.46	8.33	9.00	9.85	11.3	12.1	12.7	13.9	13.9	13.9	13.9	13.9	16.0	20.8	23.9	27.4	29.4	29.4	27.4	25.6	20.8	16.9
79	4.49	5.90	7.03	9.07	9.76	10.6	12.1	13.0	13.6	14.9	14.9	14.9	14.9	14.9	17.1	22.3	25.6	29.4	31.5	31.5	29.4	27.4	22.3	18.1
80	4.96	6.48	7.66	9.85	10.6	11.3	13.0	13.9	14.6	16.0	16.0	16.0	16.0	16.0	16.4	23.9	27.4	31.5	33.7	33.7	31.5	29.4	23.9	19.4
81	5.48	7.11	8.33	10.6	11.3	12.1	13.9	14.9	15.7	17.1	17.1	17.1	17.1	17.1	19.7	25.6	29.4	33.7	36.1	36.1	33.7	31.5	25.6	20.8
82	6.06	7.81	9.07	11.3	12.1	13.0	14.9	16.0	16.9	18.4	18.4	18.4	18.4	18.4	21.1	27.4	31.5	36.1	38.7	38.7	36.1	33.7	27.4	22.3
83	6.70	8.57	9.87	12.1	13.0	13.9	16.0	17.1	18.1	19.7	19.7	19.7	19.7	19.7	22.6	29.4	33.7	38.7	41.5	41.5	38.7	36.1	29.4	23.9
84	7.41	9.41	10.7	13.0	13.9	14.9	17.1	18.4	19.4	21.1	21.1	21.1	21.1	21.1	24.3	31.5	36.1	41.5	44.4	44.4	41.5	38.7	31.5	25.6
85	8.19	10.3	11.7	13.9	14.9	16.0	18.4	19.7	20.8	22.6	22.6	22.6	22.6	22.6	26.0	33.7	38.7	44.4	47.6	47.6	44.4	41.5	33.7	27.4
86	9.05	11.3	12.7	14.9	16.0	17.1	19.7	21.1	22.4	24.3	24.3	24.3	24.3	24.3	27.9	36.1	41.5	47.6	51.0	51.0	47.6	44.4	36.1	29.4
87	10.0	12.1	13.9	16.0	17.1	18.4	21.1	22.6	24.0	26.0	26.0	26.0	26.0	26.0	29.9	38.7	44.4	51.0	54.7	54.7	51.0	47.6	38.7	31.5
88								24.1		27.9				27.9	32.0	41.5	47.6	54.7	58.6	58.6	54.7	51.0	41.5	33.7

	1	2	3	4	5	6	7	8	9	10	11	12	13	14	15	16	17	18	19	20	21	22	23	24	25	26	27
90	13.5	14.9	16.0	17.1	18.4	19.7	21.1	22.6	24.3	26.0	27.9	32.0	32.0	32.0	32.0	32.0	32.0	36.8	47.6	54.7	62.7	67.2	67.2	62.7	58.6	47.6	38.7
91	14.9	16.0	17.1	18.4	19.7	21.1	22.6	24.3	26.0	27.9	29.9	34.3	34.3	34.3	34.3	34.3	34.3	39.4	51.0	58.6	67.2	72.0	72.0	67.2	62.7	51.0	41.5
92	16.0	17.1	18.4	19.7	21.1	22.6	24.3	26.0	27.9	29.9	32.0	36.8	36.8	36.8	36.8	36.8	36.8	42.2	54.7	62.7	72.0	77.2	77.2	72.0	67.2	54.7	44.4
93	17.1	18.4	19.7	21.1	22.6	24.3	26.0	27.9	29.9	32.0	34.3	39.4	39.4	39.4	39.4	39.4	39.4	45.3	58.6	67.2	77.2	82.7	82.7	77.2	72.0	58.6	47.6
94	18.4	19.7	21.1	22.6	24.3	26.0	27.9	29.9	32.0	34.3	36.8	42.2	42.2	42.2	42.2	42.2	42.2	48.5	62.7	72.0	82.7	88.6	88.6	82.7	77.2	62.7	51.0
95	19.7	21.1	22.6	24.3	26.0	27.9	29.9	32.0	34.3	36.8	39.4	45.3	45.3	45.3	45.3	45.3	45.3	52.0	67.2	77.2	88.6	94.9	94.9	88.6	82.7	67.2	54.7
96	21.1	22.6	24.3	26.0	27.9	29.9	32.0	34.3	36.8	39.4	42.2	48.5	48.5	48.5	48.5	48.5	48.5	55.7	72.0	82.7	94.9	102	102	94.9	88.6	72.0	58.6
97	22.6	24.3	26.0	27.9	29.9	32.0	34.3	36.8	39.4	42.2	45.3	52.0	52.0	52.0	52.0	52.0	52.0	59.7	77.2	88.6	102	109	109	102	94.9	77.2	62.7
98	24.3	26.0	27.9	29.9	32.0	34.3	36.8	39.4	42.2	45.3	48.5	55.7	55.7	55.7	55.7	55.7	55.7	64.0	82.7	94.9	109	117	117	109	102	82.7	67.2
99	26.0	27.9	29.9	32.0	34.3	36.8	39.4	42.2	45.3	48.5	52.0	59.7	59.7	59.7	59.7	59.7	59.7	68.6	88.6	102	117	125	125	117	109	88.6	72.0
100	27.9	29.9	32.0	34.3	36.8	39.4	42.2	45.3	48.5	52.0	55.7	64.0	64.0	64.0	64.0	64.0	64.0	73.5	94.9	109	125	134	134	125	117	94.9	77.2
101	29.9	32.0	34.3	36.8	39.4	42.2	45.3	48.5	52.0	55.7	59.7	68.6	68.6	68.6	68.6	68.6	68.6	78.8	102	117	134	144	144	134	125	102	82.7
102	32.0	34.3	36.8	39.4	42.2	45.3	48.5	52.0	55.7	59.7	64.0	73.5	73.5	73.5	73.5	73.5	73.5	84.4	109	125	144	154	154	144	134	109	88.6
103	34.3	36.8	39.4	42.2	45.3	48.5	52.0	55.7	59.7	64.0	68.6	78.8	78.8	78.8	78.8	78.8	78.8	90.5	117	134	154	165	165	154	144	117	94.9
104	36.8	39.4	42.2	45.3	48.5	52.0	55.7	59.7	64.0	68.6	73.5	84.4	84.4	84.4	84.4	84.4	84.4	97.0	125	144	165	177	177	165	154	125	102
105	39.4	42.2	45.3	48.5	52.0	55.7	59.7	64.0	68.6	73.5	78.8	90.5	90.5	90.5	90.5	90.5	90.5	104	134	154	177	189	189	177	165	134	109
106	42.2	45.3	48.5	52.0	55.7	59.7	64.0	68.6	73.5	78.8	84.4	97.0	97.0	97.0	97.0	97.0	97.0	111	144	165	189	203	203	189	177	144	117
107	45.3	48.5	52.0	55.7	59.7	64.0	68.6	73.5	78.8	84.4	90.5	104	104	104	104	104	104	119	154	177	203	217	217	203	189	154	125
108	48.5	52.0	55.7	59.7	64.0	68.6	73.5	78.8	84.4	90.5	97.0	111	111	111	111	111	111	128	165	189	217	233	233	217	203	165	134
109	52.0	55.7	59.7	64.0	68.6	73.5	78.8	84.4	90.5	97.0	104	119	119	119	119	119	119	137	177	203	233	249	249	233	217	177	144
110	55.7	59.7	64.0	68.6	73.5	78.8	84.4	90.5	97.0	104	111	128	128	128	128	128	128	147	189	217	249	267	267	249	233	189	154
111	59.7	64.0	68.6	73.5	78.8	84.4	90.5	97.0	104	111	119	137	137	137	137	137	137	158	203	233	267	286	286	267	249	203	165
112	64.0	68.6	73.5	78.8	84.4	90.5	97.0	104	111	119	128	147	147	147	147	147	147	169	217	249	286	307	307	286	267	217	177
113	68.6	73.5	78.8	84.4	90.5	97.0	104	111	119	128	137	158	158	158	158	158	158	181	233	267	307	329	329	307	286	233	189
114	73.5	78.8	84.4	90.5	97.0	104	111	119	128	137	147	169	169	169	169	169	169	194	249	286	329	352	352	329	307	243	203
115	78.8	84.4	90.5	97.0	104	111	119	128	137	147	158	181	181	181	181	181	181	208	267	307	352	377	377	352	329	267	217
116	84.4	90.5	97.0	104	111	119	128	137	147	158	169	194	194	194	194	194	194	223	286	329	377	404	404	377	352	286	233
117	90.5	97.0	104	111	119	128	137	147	158	169	181	208	208	208	208	208	208	239	307	352	404	433	433	404	377	307	249
118	97.0	104	111	119	128	137	147	158	169	181	194	223	223	223	223	223	223	256	329	377	433	464	464	433	404	329	267
119	104	111	119	128	137	147	158	169	181	194	208	239	239	239	239	239	239	274	352	404	464	497	497	464	433	352	286

Table 2-1. (Cont.)

SPL	50	63	80	100	125	160	200	250	315	400	500	630	800	1000	1250	1600	2000	2500	3150	4000	5000	6300	8000	10000
120	111	119	137	158	169	181	208	223	239	256	256	256	256	256	294	377	433	497	533	533	497	464	377	307
121	119	128	147	169	181	194	223	239	256	274	274	274	274	274	315	404	464	533	571	571	533	497	404	329
122	128	137	158	181	194	208	239	256	274	294	294	294	294	294	338	433	497	571	611	611	571	533	433	352
123	137	147	169	194	208	223	256	274	294	315	315	315	315	315	362	464	533	611	655	655	611	571	464	377
124	147	158	181	208	223	239	274	294	315	338	338	338	338	338	388	497	571	655	702	702	655	611	497	404
125	158	169	194	223	239	256	294	315	338	362	362	362	362	362	416	533	611	702	752	752	702	655	533	433
126	169	181	208	239	256	274	315	338	362	388	388	388	388	388	446	571	655	752	806	806	752	702	571	464
127	181	194	223	256	274	294	338	362	388	416	416	416	416	416	478	611	702	806	863	863	806	752	611	497
128	194	208	239	274	294	315	362	388	416	446	446	446	446	446	512	655	752	863	925	925	863	806	655	533
129	208	223	256	294	315	338	388	416	446	478	478	478	478	478	549	702	806	925	991	991	925	863	702	571
130	223	239	274	315	338	362	416	446	478	512	512	512	512	512	588	752	863	991	1062	1062	991	925	752	611
131	239	256	294	338	362	388	446	478	512	549	549	549	549	549	630	806	925	1062	1137	1137	1062	991	806	655
132	256	274	315	362	388	416	478	512	549	588	588	588	588	588	676	863	991	1137	1219	1219	1137	1062	863	702
133	274	294	338	388	416	446	512	549	588	630	630	630	630	630	724	925	1062	1219	1306	1306	1219	1137	925	752
134	294	315	362	416	446	478	549	588	630	676	676	676	676	676	776	991	1137	1306	1399	1399	1306	1219	991	806
135	315	338	388	446	478	512	588	630	676	724	724	724	724	724	832	1062	1219	1399	1499	1499	1399	1306	1062	863
136	338	362	416	478	512	549	630	676	724	776	776	776	776	776	891	1137	1306	1499	1606	1606	1499	1399	1137	925
137	362	388	446	512	549	588	676	724	776	832	832	832	832	832	955	1219	1399	1606	1721	1721	1606	1499	1219	991
138	388	416	478	549	588	630	724	776	832	891	891	891	891	891	1024	1306	1499	1721	1844	1844	1721	1606	1306	1062
139	416	446	512	588	630	676	776	832	891	955	955	955	955	955	1098	1399	1606	1844	1975	1975	1844	1721	1399	1137
140	446	478	549	630	676	724	832	891	955	1024	1024	1024	1024	1024	1176	1499	1721	1975			1975	1844	1499	1219
141	478	512	588	676	724	776	891	955	1024	1098	1098	1098	1098	1098	1261	1606	1844					1975	1606	1306
142	512	549	630	724	776	832	955	1024	1098	1176	1176	1176	1176	1176	1351	1721	1975						1721	1399
143	549	588	676	776	832	891	1024	1098	1176	1261	1261	1261	1261	1261	1448	1844							1844	1499
144	588	630	724	832	891	955	1098	1176	1261	1351	1351	1351	1351	1351	1552	1975							1975	1606
145	630	676	776	891	955	1024	1176	1261	1351	1448	1448	1488	1448	1448	1664									1721
146	676	724	832	955	1024	1098	1261	1351	1448	1552	1552	1552	1552	1552	1783									1844
147	724	776	891	1024	1098	1176	1351	1448	1552	1664	1664	1664	1664	1664	1911									1975
148	776	832	955	1098	1176	1261	1448	1552	1664	1783	1783	1783	1783	1783	2040									

c) the tone correction factor is added to the perceived noise level to obtain tone corrected perceived noise levels, PNLT(k), at each one-half second increment of time:

$$PNLT(k) = PNL(k) + C(k)$$

The instantaneous values of tone corrected perceived noise level are derived and the maximum value, PNLTM, is determined;

d) a duration correction factor, D, is computed by integration under the curve of tone corrected perceived noise level versus time;

e) effective perceived noise level, EPNL, is determined by the algebraic sum of the maximum tone corrected perceived noise level and the duration correction factor:

$$EPNL = PNLTM + D$$

4.2 Perceived noise level

4.2.1 Instantaneous perceived noise levels, PNL(k), shall be calculated from instantaneous one-third octave band sound pressure levels, SPL (i, k) as follows:

Step 1. Convert each one-third octave band SPL (i, k), from 50 to 10 000 Hz, to perceived noisiness n (i, k), by reference to Table 2-1, or to the mathematical formulation of the noy table given in Section 7.

Step 2. Combine the perceived noisiness values, n (i, k), found in step 1 by the following formula:

$$N(k) = n(k) + 0.15 \left\{ \left[\sum_{i=1}^{24} n(i, k) \right] - n(k) \right\}$$

$$= 0.85\, n(k) + 0.15 \sum_{i=1}^{24} n(i, k)$$

where n (k) is the largest of the 24 values of n (i, k) and N (k) is the total perceived noisiness.

Step 3. Convert the total perceived noisiness, N (k), into perceived noise level, PNL(k), by the following formula:

$$PNL(k) = 40.0 + \frac{10}{\log 2} \log N(k)$$

which is plotted in Figure 2-4. PNL(*k*) may also be obtained by choosing N(*k*) in the 1 000 Hz column of Table 2-1 and then reading the corresponding value of SPL(*i, k*) which, at 1 000 Hz, equals PNL(*k*).

4.3   Correction for spectral irregularities

4.3.1   Noise having pronounced spectral irregularities (for example, the maximum discrete frequency components or tones) shall be adjusted by the correction factor *C* (*k*) calculated as follows:

Step 1. Starting with the corrected sound pressure level in the 80 Hz one-third octave band (band number 3), calculate the changes in sound pressure level (or "slopes") in the remainder of the one-third octave bands as follows:

s (3, k) = no value

s (4, k) = SPL(4, k) − SPL(3, k)
•
•
s (i, k) = SPL(i, k) − SPL($i-1$, k)
•
•
s (24, k) = SPL(24, k) − SPL(23, k)

Step 2. Encircle the value of the slope, s (*i, k*), where the absolute value of the change in slope is greater than five; that is, where:

$$| \Delta s\ (i, k) | = | s\ (i, k) - s\ (i-1, k) | > 5$$

Step 3.

a) If the encircled value of the slope s (*i, k*), is positive and algebraically greater than the slope s (*i* − 1, *k*) encircle SPL(*i, k*).

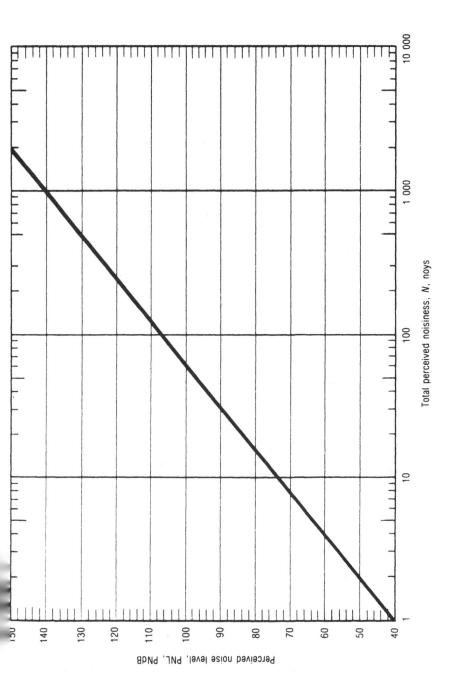

Total perceived noisiness, N, noys

Perceived noise level, PNL, PNdB

b) If the encircled value of the slope $s(i, k)$ is zero or negative and the slope $s(i-1, k)$ is positive, encircle $SPL(i-1, k)$.

c) For all other cases, no sound pressure level value is to be encircled.

Step 4. Compute new adjusted sound pressure levels $SPL'(i, k)$ as follows:

a) For non-encircled sound pressure levels, let the new sound pressure levels equal the original sound pressure levels, $SPL'(i, k) = SPL(i, k)$.

b) For encircled sound pressure levels in bands 1 to 23 inclusive, let the new sound pressure level equal the arithmetic average of the preceding and following sound pressure levels:

$$SPL'(i, k) = \tfrac{1}{2}\left[SPL(i-1, k) + SPL(i+1, k)\right]$$

c) If the sound pressure level in the highest frequency band ($i = 24$) is encircled, let the new sound pressure level in that band equal:

$$SPL'(24, k) = SPL(23, k) + s(23, k)$$

Step 5. Recompute new slope $s'(i, k)$, including one for an imaginary 25-th band, as follows:

$$s'(3, k) = s'(4, k)$$

$$s'(4, k) = SPL'(4, k) - SPL'(3, k)$$
$$\bullet$$
$$\bullet$$
$$s'(i, k) = SPL'(i, k) - SPL'(i-1, k)$$
$$\bullet$$
$$\bullet$$
$$s'(24, k) = SPL'(24, k) - SPL'(23, k)$$

$$s'(25, k) = s'(24, k)$$

Step 6. For i from 3 to 23 compute the arithmetic average of the three adjacent slopes as follows:

Step 7. Compute final one-third octave-band sound pressure levels, SPL"(i, k), by beginning with band number 3 and proceeding to band number 24 as follows:

SPL"$(3, k)$ = SPL$(3, k)$

SPL"$(4, k)$ = SPL"$(3, k)$ + $\bar{s}(3, k)$

•
•

SPL"(i, k) = SPL"$(i - 1, k)$ + $\bar{s}(i - 1, k)$

•
•

SPL"$(24, k)$ = SPL"$(23, k)$ + $\bar{s}(23, k)$

Step 8. Calculate the differences, $F(i, k)$ between the original sound pressure level and the final background sound pressure level as follows:

$$F(i, k) = \text{SPL}(i, k) - \text{SPL}''(i, k)$$

and note only values equal to or greater than one and a half.

Step 9. For each of the relevant one-third octave bands (3 to 24), determine tone correction factors from the sound pressure level differences $F(i, k)$ and Table 2-2.

Step 10. Designate the largest of the tone correction factors, determined in Step 9, as $C(k)$. An example of the tone correction procedure is given in Table 2-3.

Tone corrected perceived noise levels PNLT(k) shall be determined by adding the $C(k)$ values to corresponding PNL(k) values, that is:

$$\text{PNLT}(k) = \text{PNL}(k) + C(k)$$

For any i-th one-third octave band, at any k-th increment of time, for which the tone correction factor is suspected to result from something other than (or in addition to) an actual tone (or any spectral irregularity other than aeroplane noise), an additional analysis shall be made using a filter with a bandwidth narrower than one-third of an octave. If the narrow band analysis corroborates these suspicions, then a revised value for the background sound pressure level SPL"(i, k), shall be determined from the narrow band analysis and used to compute a revised tone correction factor for that particular one-third octave band.

Note.— Other methods of rejecting spurious tone corrections such as those described in 2.1.5.1 a), 3.1.4 and Appendix 2 of the Environmental Technical Manual on the use of Procedures in the Noise Certification of Aircraft *(Doc 9501) may be used.*

4.3.2   This procedure will underestimate EPNL if an important tone is of a frequency such that it is recorded in two adjacent one-third octave bands. It shall be

Table 2-2.   Tone Correction Factors

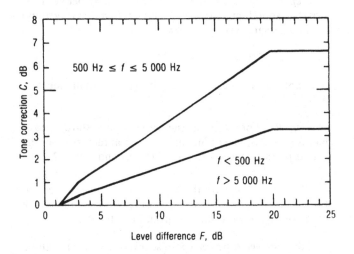

Frequency f, Hz	Level difference F, dB	Tone correction C, dB
$50 \leq f < 500$	$1\tfrac{1}{2}* \leq F < 3$	$F/3 - \tfrac{1}{2}$
	$3 \leq F < 20$	$F/6$
	$20 \leq F$	$3\tfrac{1}{3}$
$500 \leq f \leq 5\,000$	$1\tfrac{1}{2}* \leq F < 3$	$2\,F/3 - 1$
	$3 \leq F < 20$	$F/3$
	$20 \leq F$	$6\tfrac{2}{3}$
$5\,000 < f \leq 10\,000$	$1\tfrac{1}{2}* \leq F < 3$	$F/3 - \tfrac{1}{2}$
	$3 \leq F < 20$	$F/6$
	$20 \leq F$	$3\tfrac{1}{3}$

* See Step 8, 4.3.1.

demonstrated to the satisfaction of the certificating authority:

either that this has not occurred,

or that if it has occurred that the tone correction has been adjusted to the value it would have had if the tone had been recorded fully in a single one-third octave band.

4.4 Maximum tone corrected perceived noise level

4.4.1 The maximum tone corrected perceived noise level, PNLTM, shall be the maximum calculated value of the tone corrected perceived noise level PNLT(k). It shall be calculated in accordance with the procedure of 4.3. To obtain a satisfactory noise time history, measurements shall be made at 500 ms time intervals.

Note.— Figure 2-5 is an example of a flyover noise time history where the maximum value is clearly indicated.

4.4.2 If there are no pronounced irregularities in the spectrum, even when examined by a narrow-band analysis, then the procedure of 4.3 shall be disregarded since PNLT(k) would be identically equal to PNL(k). For this case, PNLTM shall be the maximum value of PNL(k) and would equal PNLM.

4.5 Duration correction

4.5.1 The duration correction factor D determined by the integration technique shall be defined by the expression:

$$D = 10 \log \left[\left(\frac{1}{T} \right) \int_{t(1)}^{t(2)} \text{antilog } \frac{\text{PNLT}}{10} \, dt \right] - \text{PNLTM}$$

where T is a normalizing time constant, PNLTM is the maximum value of PNLT, $t(1)$ is the first point of time after which PNLT becomes greater than PNLTM $-$ 10 and $t(2)$ is the point of time after which PNLT remains constantly less than PNLTM $-$ 10.

Table 2-3. Example of tone correction calculation for a turbofan engine

① Band (i)	② f Hz	③ SPL dB	④ S dB Step 1	⑤ 1ΔS1 dB Step 2	⑥ SPL' dB Step 4	⑦ S' dB Step 5	⑧ $\bar{S}$ dB Step 6	⑨ SPL'' dB Step 7	⑩ F dB Step 8	⑪ C dB Step 9
1	50	—	—	—	—	—	—	—	—	—
2	63	—	—	—	—	—	—	—	—	—
3	80	70	—	—	70	-8	$-2\frac{1}{3}$	70	—	
4	100	62	-8	—	62	-8	$+3\frac{1}{3}$	$67\frac{2}{3}$	—	
5	125	70	$+8$	16	71	$+9$	$+6\frac{2}{3}$	71	—	
6	160	80	$+10$	2	80	$+9$	$+2\frac{2}{3}$	$77\frac{2}{3}$	$2\frac{1}{3}$	0.39
7	200	82	$+2$	8	82	$+2$	$-1\frac{1}{3}$	$80\frac{1}{3}$	$1\frac{2}{3}$	0.28
8	250	83	$+1$	1	79	-3	$-1\frac{1}{3}$	79	4	0.66
9	315	76	-7	8	76	-3	$+\frac{1}{3}$	$77\frac{2}{3}$	—	
10	400	80	$+4$	11	78	$+2$	$+1$	78	2	0.33
11	500	80	0	4	80	$+2$	0	79	1	0.33
12	630	79	-1	1	79	-1	0	79	—	
13	800	78	-1	0	78	-1	$-\frac{1}{3}$	79	—	
14	1 000	80	$+2$	3	80	$+2$	$-\frac{2}{3}$	$78\frac{2}{3}$	$1\frac{1}{3}$	0.44
15	1 250	78	-2	4	78	-2	$-\frac{1}{3}$	78	—	

16	1 600	76	− 2	0	76	− 2	$+\frac{1}{3}$	$77\frac{2}{3}$	—	
17	2 000	79	+ 3	5	79	+ 3	+1	78	1	0.33
18	2 500	(85)	+ 6	3	79	0	$-\frac{1}{3}$	79	6	[2]
19	3 150	79	−(6)	12	79	0	$-2\frac{2}{3}$	$78\frac{2}{3}$	$\frac{1}{3}$	0.11
20	4 000	78	− 1	5	78	− 1	$-6\frac{1}{3}$	76	2	0.66
21	5 000	71	−(7)	6	71	− 7	− 8	$69\frac{2}{3}$	$1\frac{1}{3}$	0.44
22	6 300	60	−11	4	60	−11	$-8\frac{2}{3}$	$61\frac{2}{3}$	—	
23	8 000	54	− 6	5	54	− 6	− 8	53	1	0.16
24	10 000	45	− 9	3	45	− 9	−	45	—	
						− 9				

Step 1	③(i) − ③(i−1)
Step 2	|④(i) − ④(i−1)|
Step 3	see instructions
Step 4	see instructions
Step 5	⑥(i) − ⑥(i−1)

Step 6	[⑦(i)+⑦(i+1)+ ⑦(i+2)]÷3
Step 7	⑨(i−1)+⑧(i−1)
Step 8	③(i) − ⑨(i)
Step 9	see Table 2-2

Note.— Steps 5 and 6 may be eliminated in the calculations if desired. In this case in the example shown in Table 2-3, columns ⑦ and ⑧ should be removed and existing columns ⑨, ⑩ and ⑪ become ⑦, ⑧ and ⑨ covering new steps 5, 6 and 7 respectively. The existing steps 5, 6, 7, 8 and 9 in 4.3.1 are then replaced by:

STEP 5 [⑥(i − 1) + ⑥ i + ⑥(i + 1)] ÷ 3

STEP 6 ③(i) − ⑦(i) if > 0

STEP 7 See Table 2-2.

Figure 2-5. Example of perceived noise level corrected for tones as a function of aeroplane flyover time

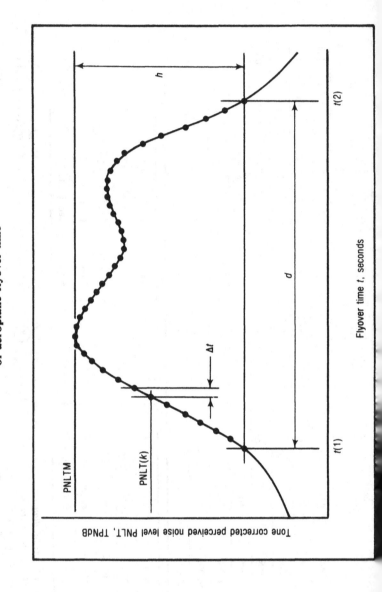

4.5.2 Since PNLT is calculated from measured values of SPL, there will, in general, be no obvious equation for PNLT as a function of time. Consequently, the equation shall be rewritten with a summation sign instead of an integral sign as follows:

$$D = 10 \log \left[\left(\frac{1}{T} \right) \sum_{k=0}^{d/\Delta t} \Delta t \cdot \text{antilog} \frac{\text{PNLT}(k)}{10} \right] - \text{PNLTM}$$

where Δt is the length of the equal increments of time for which PNLT(k) is calculated and d is the time interval to the nearest 1.0 s during which PNLT(k) remains greater or equal to PNLTM − 10.

4.5.3 To obtain a satisfactory history of the perceived noise level,

a) half-second time intervals for Δt, or

b) a shorter time interval with approved limits and constants,

shall be used.

4.5.4 The following values for T and Δt shall be used in calculating D in the procedure given in 4.5.2:

$$T = 10 \text{ s, and}$$
$$\Delta t = 0.5 \text{ s}$$

Using the above values, the equation for D becomes

$$D = 10 \log \left[\sum_{k=0}^{2d} \text{antilog} \frac{\text{PNLT}(k)}{10} \right] - \text{PNLTM} - 13$$

where the integer d is the duration time defined by the points corresponding to the values PNLTM − 10.

4.5.5 If in the procedures given in 4.5.2, the limits of PNLTM − 10 fall between the calculated PNLT(k) values (the usual case), the PNLT(k) values defining the limits of the duration interval shall be chosen from the PNLT(k) values closest to PNLTM − 10.

4.6 Effective perceived noise level

4.6.1 The total subjective effect of an aeroplane flyover, designated effective perceived noise level, EPNL, shall be equal to the algebraic sum of the maximum value of the tone corrected perceived noise level, PNLTM, and the duration correction D. That is:

$$EPNL = PNLTM + D$$

where PNLTM and D are calculated in accordance with the procedures given in 4.2, 4.3, 4.4 and 4.5.

APPENDIX 5

Calculation of aircraft noise contours around airports

Noise contours, or statements of the noise level heard at various positions on the ground around an airport, are computed rather than measured because of the large areas of ground covered and the length of time over which noise data have to be averaged.

The noise at any point on the ground in the vicinity of an aircraft operation will depend upon a number of factors. Uppermost amongst these are the types of aircraft and their power-plant; the power, flap setting and airspeed conditions throughout the operation; the distances from the points on the ground to the aircraft; and the effect of local topography and weather on propagation of the noise. Most airport operations will include a variety of aircraft and flight procedures and a wide range of operational weights. Because of the large quantity of data (which can be regarded as aircraft or airport specific) required to compute the noise of each individual operation, it is normal to make certain simplifications to reflect average noise exposure over long time periods. These time periods may range from as little as one day to several months.

The normal process of computation embraces three main steps:

1. the calculation of noise level from individual aircraft movements at a matrix of observation points around the airport;
2. the integration of all the individual noise levels over a defined period of time; and
3. the interpolation and plotting of the information in the form of noise contours.

The simplifying assumptions that are most frequently made include the noise levels of "groups" of similar aircraft types, average climatic conditions and the average operational pattern over the time period in question. Given the necessary computer power and dedication, it is possible to produce contours from a far more detailed study of individual operations, but this is unusual. Either way, the following information is required to allow the contour calculation process to be undertaken:

aircraft types that use the airport;

noise–power–distance relationships for each type;

aircraft performance data for each type;

routes used in departure and arrival;

number of movements on each route within the period chosen;

operational data typical to each route, including aircraft mass, power setting, speed and configuration through the different flight segments; and

airport-related data, including meteorological conditions and physical alignment of runways.

The airport operator will normally provide information on aircraft types and the number of movements. The aircraft manufacturer will be the source of noise and performance data, and the airlines or operators are the best source for operational and flight data. Arrival and departure routings are normally determined by the Air Traffic Control Authority.

As discussed in Section 7.2, there has now been considerable standardisation in the methods of deriving noise and performance data, as well as the general effects that influence noise projections, and in contour calculation processes. The International Civil Aviation Organisation (ICAO) has published a recommended method[990] for computing noise contours around airports, in which the following critical elements are discussed:

A5.1   Noise–power–distance information

Noise–power–distance data usually take the form illustrated in Figure A.5.1. They are based on measurements taken under defined conditions (usually certification tests) at a particular distance, but are extrapolated by the manufacturer to cover a range of distances under sound propagation conditions that have been determined as the average of those occurring at a representative sample of major world airports. Note, these are not the same conditions as are assumed in noise certification. The noise–power–distance curves cover a range of relevant power settings, with defined aircraft configurations as used during normal take-off and approach procedures. These will usually be sufficient to cover the individual segments in a total operation (illustrated in Fig. A.5.2) and will allow noise calculation at all sound-pressure levels that are relevant in the community. There is normally a lower-level cut-off in the noise data, higher than the urban background level but no higher than what would be considered as the threshold of community annoyance.

A5.2   Aircraft performance data

Reference flight profiles, associated engine thrust information and aircraft speeds are normally required from the aircraft manufacturers of a given aircraft type when it undertakes two typical operations. These embrace a take-off at 85% of maximum take-off mass, including a noise abatement cut-back procedure to minimise community disturbance; and a 3° approach profile for which the thrust is that needed to land the aircraft at 90% of maximum landing mass.

In addition, coefficients relating aircraft performance to altitude, tempera-

Figure A.5.1. Noise – power – distance curves for an individual aircraft.

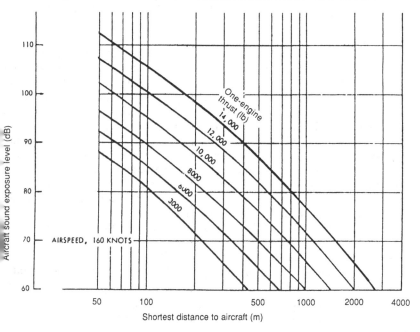

Figure A.5.1. Noise–power–distance curves for an individual aircraft.

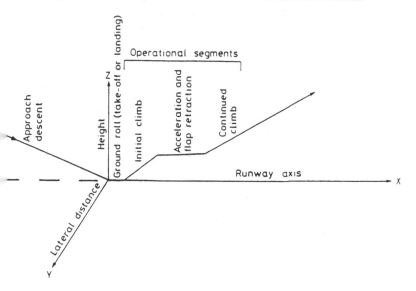

ture, wind, aeroplane weight and engine thrust are provided to allow other flight profiles to be computed as differences from the reference profiles.

A5.3 Calculation method

For each aircraft movement (arrival or departure), aircraft position information and engine thrust details are computed throughout the various operational segments. At any selected point on the ground (J in Fig. A.5.3) the shortest distance to the flight path is calculated and the noise data are interpolated for that distance (d) and the engine thrust. The aircraft position information has to allow for some lateral displacement of the actual ground track from the nominal to account for the natural dispersion (variability) that takes place in everyday operations. Corrections also have to be applied for overground and other lateral attenuation effects to account for the natural reduction of noise to the side of the take-off run and the early part of the climb, and for directivity characteristics before the start of the take-off run. Other corrections also need to be applied throughout the flight envelope for aircraft speed changes and for changes in noise duration that occur during an aircraft turn (e.g., the noise stays at a constant level at the centre of radius of a turn, but has a much shorter duration on the outside of the turn radius).

Hence, the noise level at any point J arising from an individual aircraft movement is expressed by the following formula:

$$L_J = L(X,d) + \Delta(\beta,l) + \Delta_D(Q) + \Delta_S(V) + \Delta_T(T),$$

where

$L(X,d)$ = noise level interpolated from the noise-thrust-distance data,

$\Delta(\beta,l)$ = extra ground attenuation (a function of the elevation angle β and distance to the ground track l),

$\Delta_D(Q)$ = correction for directivity behind the start of roll (a function of the angle Q subtended to the rear of the aircraft),

Figure A.5.3. Geometry of an observation point and an aircraft flight path.

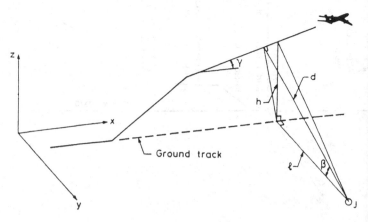

$\Delta_S(V)$ = correction for aeroplane speed V, and
$\Delta_T(T)$ = correction for changes in duration during turns, a function of time T.

To represent a pattern of operations at an airport, the above calculation is repeated for the same point on the ground for all movements of various aircraft during the time period of interest. The whole process is repeated at a sufficient number of other points on the ground to permit contours to be plotted. The type of contour plot that will emerge from an automated computer output system is shown in Figure A.5.4. This can then be related to the map of the local district.

For ease of computation, it might not be practical to account for each movement, and there may well be a case for "grouping" aircraft of the same type, or similar types having almost the same noise characteristics.

A5.4 Current shortcomings

Although there has been general agreement on the contour calculation process, even the combined efforts of ICAO, the SAE and ECAC have failed to produce a complete suite of high-reliability subroutines to the total programme. Those that are deficient in some respect include the way in which the take-off ground-roll phase of operation is treated and the "broad-brush" relationship for lateral attenuation. Equally, although the method does notionally apply to propeller-powered aircraft as well as to jet- and turbofan-powered aircraft, it does not apply to helicopters, owing to the unique directivity patterns of their noise sources and the particular difficulties of specifying flight patterns. There is also some doubt as to the credibility of the method where airfields are used purely for general aviation purposes, because of the

Figure A.5.4. A typical set of airport noise contours obtained from computer printout.

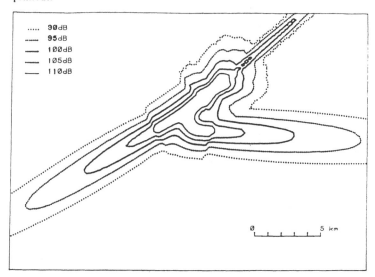

irregularity of flight patterns and because of the difficulty and cost of gathering the necessary background noise and performance data on the aircraft types concerned.

It should be recognised that noise contours are primarily intended to apply at commercial airports, where the bulk of operations are from jet- and turbofan-powered aircraft, and should really be restricted to determining changes in the long-term average conditions, not absolute contour dimensions or boundaries between the acceptable and the unacceptable.

APPENDIX 6

Typical aircraft noise levels

Data are summarised in Table A.6.1. and Figure A.6.1.

Table A.6.1. *Aircraft noise levels at certification measurement points*

Aircraft type	450-m sideline (EPNL)	6.5-km take-off (EPNL)	(dBA)	2-km approach (EPNL)	(dBA)
Boeing 707	115	114	104	118	105
Boeing 727	102	101	88	104	91
Boeing 737	101	96	87	102	92
Boeing 747	101	105	96	105	97
Boeing 757	94	89	71	97	86
Boeing 767	96	90	74	102	89
Douglas DC8	114	114	102	117	104
Douglas DC8-70	93	95	85	99	88
Douglas DC9	102	97	87	102	90
Douglas DC10/MD11	98	100	90	106	94
Douglas MD80	96	90	82	93	84
Lockheed L1011	96	98	86	102	91
Airbus A300	96	91	78	102	91
Airbus A310	97	89	76	100	89
Airbus A320	93	85	72	92	81
BAe Trident	106	105	95	105	95
BAe 1-11	103	96	88	102	92
BAe 146	88	85	74	96	86
Fokker F28	100	93	79	101	93
Fokker F100	89	84	72	93	83
Concorde	119	119	113	116	109
Old business jets	102	100	85	105	88
Gulfstream 4	86	79	67	91	81

Sources: ICAO and FAA published measured or estimated data for certification conditions. Figures quoted are typical of popular production models, and are for guidance only. For definitive data consult manufacturers or referenced documentation.[35]

Figure A.6.1. Typical aircraft noise level (ICAO Annex 16, Chapter 3 rules).

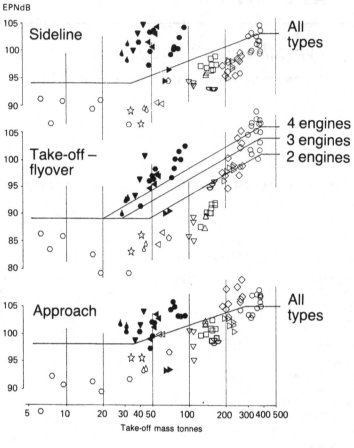

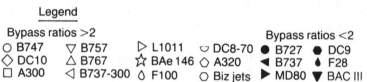

Legend

Bypass ratios >2

○ B747	▽ B757	▷ L1011	⌣ DC8-70	
◇ DC10	△ B767	☆ BAe 146	◇ A320	
□ A300	◁ B737-300	◊ F100	○ Biz jets	

Bypass ratios <2

● B727	● DC9
◀ B737	◖ F28
▶ MD80	▼ BAC III

326

References

The references in this book do not appear in numerical order. The reason for this is simply that all the references have been categorised by subject matter, and appear within the following ten broad topics:

Legal, operational and general

1. Anon. (1988). "International Standards and Recommended Practices – Environmental Protection – Annex 16 to the Convention on International Civil Aviation; Volume 1, Aircraft Noise. Second Edition – 1988" (first published simply as "Aircraft Noise" in 1971 and then reissued in 1975, 1977 and 1981). International Civil Aviation Organisation (ICAO), Montreal.
2. Anon. (1988). "ICAO Environmental Technical Manual on the Use of Procedures in the Noise Certification of Aircraft". ICAO Doc. 9501-AN/929.
3. Anon. (1969 onwards). Reports of the 1969 Special ICAO Meeting and subsequent seven meetings of the Committee on Aircraft Noise (CAN/1, /2, /3 et seq.), held between 1970 and 1983; and the 1st meeting of the Committee on Aviation Environmental Protection (CAEP) in 1986.
4. Blatt, J. D. (1 September 1965). Letter from Federal Aviation Agency (now Federal Aviation Administration) to United States (U.S.) industry proposing federal action to control aircraft noise.
5. Anon. (1969 onwards). "Federal Aviation Regulations (FAR), Part 36. Noise Standards: Aircraft Type and Airworthiness Certification". U.S. Department of Transportation (DOT); Federal Aviation Administration (FAA), Washington D.C.
6. Anon. "Noise Certification Test and Analysis Procedures". FAA Advisory Circular AC 36-4 (and updates A, B, C et seq.).
7. Anon. "British Civil Airworthiness Requirements; Section N, Noise". Civil Aviation Authority (CAA), London.
8. Anon. (1981). "Subpart E – Operating Noise Limits" (and subsequent amendments). Part 91 of U.S. Federal Aviation Regulations, FAA.
9. Anon. (1979 and 83). "Council Directive 80/51/EEC on the Limitation of Noise Emissions from Subsonic Aircraft" and Amendment 83/206. Council of the European Communities (EC).

References   329

10. Anon. (1985). "Amended Commercial Airline Access Plan and Regulation, 1st April 1985 – 31st March 1990". John Wayne Airport, Orange County, Calif., USA.
11. Anon. (1970). "Noise Standards" (and proposed Amendment of 29 September 1987). California Department of Aeronautics, Administrative Code Register 70, No. 48, Chapter 9.
12. Anon. (1981). "National Capital Aiports". New Paragraph 159.40 in Federal Aviation Regulations, Part 159. FAA; Federal Register, 27 November 1981.
13. Anon. (1974). "Opposition to Two-Segment Approach Surprises Agencies". Aviation Daily, 31 July 1974.
14. Arde, Inc. and Town and City, Inc. (1966). "Study of Optimum Use of Land Exposed to Aircraft Landing and Takeoff Noise". U.S. National Aeronautics and Space Administration (NASA), Contractor Report (CR) 410.
15. Anon. (1986). "Landing Fees Provide over $2 million in Aid to Paris Airport Neighbours". Noise Regulation Reporter, U.S. Bureau of National Affairs, September 1986.
16. Anon. (1973). "Noise Insulation Regulations". U.K. Department of the Environment (DoE), HMSO, London.
17. Owen, Kenneth. (1982). "Concorde. A New Shape in the Sky". Jane's Publishing Company Ltd., London.
18. Hooker, S. G. (1984). "Not Much of an Engineer: An Autobiography". Airlife Publications.
19. Coleman, W. T., Jr. (1976). "The Secretary's Decision on Concorde Supersonic Transport". U.S. DOT.
20. Smith, M. J. T. (Chairman) (1978). "Report of Working Group E (to ICAO Committee on Aircraft Noise) on SST Noise Prediction Methods". Final report, 15 June 1978; Interim Report, 21 March 1978. ICAO.
21. Graf, D. A. (1987). "Hypersonic Civil Transports Environmental and Economic Feasibility". American Institute of Aeronautics and Astronautics (AIAA), Paper AIAA 87-2952.
22. Morris, S. J., Jr., and Weidner, J. P. (1987). "Propulsion System Issues for the High-Speed Civil Transport Study". AIAA 87-2938.
23. Welge, H. R., and Graf, D. A. (1987). "Vehicle Concept Design Consideration for Future High-Speed Commercial Flight". AIAA 87-2927.
24. Calder, P. H., and Gupta, P. C. (1975). "Future SST Engines, with Particular Reference to the Olympus-593 Evaluation and Concorde Experience". Society of Automotive Engineers (SAE) 751056.
25. Dusa, D. J. (1987). "High Mach Propulsion System Installation and Exhaust System Design Considerations". AIAA 87-2941.
26. Hay, J. A. (1976). "Concorde Community Noise". SAE 760898.
27. Pianko, M. (1975). "Methodology of Aircraft (Noise) Nuisance Certification". L'Aeronautique et l'Astronautique, No. 54.
28. Smith, M. J. T. (1982). "The Impact and Future Direction of Aircraft Noise Certification". Noise Control Engineering 18 (2).
29. SKS Group Ltd. (1985). "Transcript of Proceedings: FAA Public Hearing, 10 Dec 1985" (held in response to House of Representatives Report 99-256, instructing the FAA to report on ways of modernising the U.S. fleet). FAA.
30. Anon. (1986). "Alternative Available to Accelerate Commercial Aircraft Fleet Modernisation". FAA report to U.S. Senate and House Appropriations Committees.
31. Reilly, J. D. (Ch) (1987). "Report of the Working Group on Aircraft Noise/Airport Capacity". Industry Task Force Letter and Report to FAA Administrator.
32. Strang, W. J., and McKinlay, R. H. (1978). "Concorde in Service". U.K. Institution of Mechanical Engineers (I. Mech. E.), Proc 192 (38).
33. Anon. (1980). "Airport Environmental Handbook". FAA.

330   *References*

34. Anon. (1978). "Harmony with Aviation – The Environmental Process". U.S. DOT.
35. Anon. "Advisory Circulars, Airplane Noise Levels, Estimated and Certificated".
    AC-36-1, 2 and 3 (and/or updates 1A, B, et seq.). FAA.
36. Anon. (1987). "Review of Night Restrictions Policy" (at Heathrow and Gatwick
    Airports). Consultation Papers, U.K. Dept of Transport (DTp).
37. Miller, S. C., and Bennett, H. (1986). "Future Trends in Propulsion". Proc. ICAS
    Vol. 1, pp. 25–40.
38. Eckardt, D. (1986). "Technical/Economic Evaluation of New Propfan Concepts in
    Comparison with the Turbofan of the 1990s" (in German). BMFT, Statusseminar
    ueber Luftfahrtforschung und Luftfahrttechnologie, Munich.)
39. Miller, S. C. (1987). "Future Trends – A European View". AIAA Proc. ISABE
    Paper 87-2941.
40. Sadler, J. H. R. (1987). "Trends in Civil Aircraft Propulsion". I. Mech. Eng.
    Aerospace Technology Conference, Birmingham, England.
41. Borradaile, J. A. (1988). "Towards the Optimum Ultra High Bypass Ratio (UHBR)
    Engine". AIAA-88-2954.
42. Brathney, J. A. (Feb. 1985). "Is the Unducted Fan Tomorrow's Subsonic Engine?"
    Aeropace Engineering.
43. Anon. (1988). "Resurgence of the Prop". Aviation Week and Space Technology
    (AWST), 25 January 1988.
44. Smith, M. J. T., and Szewczyk, V. M. (1988). "What Should Be Done with Those
    Noisy Old Aircraft?" European Noise Legislation Conference, London, 2 September
    1988. Avmark International Ltd.
45. Anon. (1984). "Information on Levels of Environmental Noise Requisite to Protect
    Public Health and Welfare with an Adequate Margin of Safety". U.S. Environmental
    Protection Agency (EPA).
46. Anon. (1987). "International Air Transport: Building for Growth". AWST, 9
    November 1987.
47. Smith, M. J. T, Lowrie, B. W., Brooks, J. R., and Bushell, K. W. (1988). "Future
    SST Noise – Lessons from the Past". AIAA 88-2989.
48. Driver, C. (1987). "Report of Working Group on Aircraft Research and
    Development (Battelle Institute, Columbus, Ohio)". Eagle Inc.
49. Sutton, O. G. (1955). "The Science of Flight". Penguin Books Ltd.
50. Heller, H. (1982). "Propeller Aircraft Noise Certification and Flight Testing".
    DFVLR Report Mitt. 82.16.
51. Strack, W. C., and Morris, S. J. (1988). "The Challenges and Opportunities of
    Supersonic Transport Propulsion Technology". AIAA-88-2985.
52. Powell, R. D. (1989). "Potential Liability for Federal Preemption of Noise In and
    Around Airports – Exploding the Myth". American Bar Association; Forum
    Committee on Air and Space, February 1989.

Standards, units, noise impact and rating

100. Anon. "Octave, half-octave and third-octave band filters intended for the analysis
     of sounds and vibrations". Commission Electrotechnique Internationale (IEC),
     Geneva. IEC 225.
101. Zwicker, E. (1961). "Subdivision of the Audible Frequency Range into Critical
     Bands (Frequensgruppen)". Journal of the Acoustical Society of America (J. Ac. Soc.
     Am.) 33:248.
102. Anon. "Preferred Frequencies for Acoustical Measurements". International
     Organisation for Standards (ISO), Geneva. ISO 266.
103. Anon. "Frequency Weighting for the Measurement of Aircraft Noise". IEC 537.
104. Fletcher, H., and Munson, W. A. (1933). "Loudness, Its Definition, Measurement
     and Calculation". J. Ac. Soc. Am. 5:82.

105. Kryter, K. D. (1970). "The Effects of Noise on Man". Academic Press, New York.
106. Kryter, K. D. (1984). "Physiological, Psychological and Social Effects of Noise". NASA Publication 1115.
107. Anon. "Noise Assessment with Respect to Community Response". ISO R1996.
108. Anon. (1977). "Impact of Noise on People". DOT.
109. Ollerhead, J. B. (1978). "Predicting Public Reaction to Noise from Mixed Sources". Inter-noise 78:579–84.
110. Hall, F. L. (1982). "Community Response to Noise: Is All Noise the Same?" J. Ac. Soc. Am. 70(6).
111. Newman, J. S., and Beattie, K. R. (1985). "Aviation Noise Effects". FAA Report No. FAA-EE 85-2.
112. Rylander, R., and Sorensen, S. (1972). "Annoyance Reactions from Aircraft Noise Exposure". Journal of Sound and Vibration (J. S. Vib.) 24(4):419–44.
113. Kryter, K. D. (1959). "Scaling Human Reactions to the Sound from Aircraft". J. Ac. Soc. Am. 31:415–29.
114. Finke, H. O., Martin, R., Guski, R., Rohrmann, B., Schumer, R., and Schumer-Kohrs, A. (1982). "Effects of Aircraft Noise on Man". J. Ac. Soc. Am. 72:1222–42.
115. Ollerhead, J. B. (1971). "An Evaluation of Methods for Scaling Aircraft Noise Perception". NASA CR-1883.
116. Robinson, D. W. (1970). "The Assessment of Noise, With Particular Reference to Aircraft". Royal Aeronautical Society (RAeS) Journal (April):147–60.
117. Sorensen, S., Berglund, K., and Rylander, R. (1973). "Reaction Patterns in Annoyance Response to Aircraft Noise". Proc. Int. Conf. Noise as a Public Health Problem, Dubrovnik.
118. Galloway, W. J. (1974). "Community Noise Exposure Resulting from Aircraft Operations: Technical Review". AMRC-TR-73-106.
119. Richards, E. J., and Ollerhead, J. B. (1973). "Noise Burden Factor – New Way of Rating Airport Noise". J. S. Vib. 7(12).
120. Galloway, W. J., and Bishop, D. E. (1970). "Noise Exposure Forecasts: Evolution, Evaluation, Extensions and Land Use Interpretations". FAA Report 70-9.
121. Pearson, K. (1974). "Handbook of Noise Ratings". NASA CR-2376.
122. Anon. "Acoustics – Expression of Physical and Subjective Magnitudes of Sound or Noise in Air". ISO R131.
123. Anon. "Normal Equal Loudness Contours for Pure Tones and Threshold of Hearing under Free-Field Conditions". ISO R226.
124. Anon. "Method for Calculating Loudness Level". ISO 532.
125. Stevens, S. S. (1961). "Procedure for Calculating Loudness; Mk. VI." J. Ac. Soc. Am. 33:1577.
126. Robinson, D. W., and Dadson, R. S. (1956). "A Re-determination of the Equal Loudness Relation for Pure Tones". British Journal of Applied Physics (B. J. App. Phys.) 7:166.
27. Kryter, K. D. (1968). "Concepts of Perceived Noisiness, Their Implementation and Application". J. Ac. Soc. Am. 43:344.
28. Young, R. W., and Peterson, A. (1969). "On Estimating Noisiness of Aircraft Sounds". J. Ac. Soc. Am. 45:834.
29. Berglund, B., Berglund, U., and Lindvall, T. (1975). "Scaling Loudness, Noisiness and Annoyance of Aircraft Noise". J. Ac. Soc. Am. 57:930–4.
30. Kryter, K. D., and Pearsons, K. S. (1964). "Modification of Noy Tables". J. Ac. Soc. Am. 36:394.
31. Kryter, K. D. (1969). "Possible Modifications to Procedures for the Calculation of Perceived Noisiness". NASA Report CR-1635.
32. Anon. "Definitions and Procedures for Computing the Perceived Noise Level of Aircraft Noise". SAE ARP 865A

133. Anon. "Definitions and Procedures for Computing the Effective Perceived Noise Level for Aircraft Flyover Noise". SAE ARP 1071.
134. Anon. "Frequency Weighting Network for Approximation of Perceived Noise Level for Aircraft Noise". SAE ARP 1080.
135. Kryter, K. D., and Pearsons, K. S. (1963). "Some Effects of Spectral Content and Duration on Perceived Noise Level". J. Ac. Soc. Am. 35:886.
136. Kryter, K. D. (1966). "Review of Methods for Measuring the Loudness and Noisiness of Complex Sounds". NASA CR-422.
137. Pearsons, K. S., Horonjeff, R. D., and Bishop, D. E. (1968). "Noisiness of Tones Plus Noise". NASA CR-1117.
138. Wells, R. J. (1969). "Calculation of an Annoyance Level for Sounds Containing Multiple Pure Tones". 78th Ac. Soc. Am. Mtg.
139. Anon. (1966). "The Effect of Duration and Background Noise Level on Perceived Noisiness". U.S. FAA ADS-78.
140. Bullen, R. B., and Hede, A. J. (1983). "Time of Day Corrections in Measures of Aircraft Noise Exposure". J. Ac. Soc. Am. 73:1624–30.
141. Schultz, J. (1972). "Community Noise Rating". Applied Acoustics Supplement 1.
142. Burck, W., and Grutzmacher, M. (1965). "Aircraft Noise; Its measurement and Evaluation, Its Significance for Community Planning, and Measures for Its Abatement". Report for the German Federal Ministry of Health, Gottingen.
143. Robinson, D. W. (1971). "Towards a Unified System of Noise Assessment" J. S. Vib. 14:279. (See also NPL Aero Report Ac. 38, March 1969.)
144. Davies, L. I. C. (1977). "A Guide to the Calculation of NNI". Civil Aviation Authority (CAA), U.K. Directorate of Operational Research and Analysis (DORA), Report DORA 7908, 2d edition.
145. Wilson, A. H. (Ch.) (1963). "Noise – Final Report of the Committee on the Problem of Noise". Her Majesty's Stationery Office (HMSO), London. (Reprinted, 1973.)
146. McKennel, A. C. (1963). "Aircraft Noise Annoyance around London (Heathrow) Airport: The Social Survey". Central Office of Information (COI), London. Report 55337.
147. Rice, C. G. (1977). "Investigation of the Trade-off Effects of Aircraft Noise and Number". J. S. Vib. 52:325–44.
148. Rice, C. G. (1978). "Trade-off Effects of Aircraft Noise and Number". Proc. Intern. Commission of Biological Effects of Noise, Frieberg, Germany.
149. Powell, C. A. (1980). "Annoyance due to Multiple Airplane Noise Exposure". NASA TP-1706.
150. Rylander, R., Bjorkman, M., Ahrlin, U., Sorensen, S., and Berglund, K. (1980). "Aircraft Noise Annoyance Contours: Importance of Overflight Frequency and Noise Level". J. S. Vib. 69:586.
151. Anon. (1971). "Second Survey of Aircraft Annoyance around London (Heathrow) Airport". MIL Research Limited, HMSO.
152. Brooker, P., Critchley, J. B., Monkman, D. J. and Richmond, C. (1985). "United Kingdom Aircraft Noise Index Study : Main Report". CAA Report DR 8402.
153. Rice, C. G. (1982). "A Synthesis of Studies on Noise-Induced Sleep Disturbance". Institute of Sound and Vibration Research (ISVR), Memorandum 623, University of Southampton.
154. Borsky, P. N. (1976). "Sleep Interference and Annoyance by Aircraft Noise". Sound and Vibration, 18–21 December.
155. Ollerhead, J. B. (1977). "Variation of Community Noise Sensitivity with Time of Day". Inter-noise 77:B692–7.
156. Anon. (1980). "Aircraft Noise and Sleep Disturbance". CAA (DORA), Final Report 8008.

157. Anon. (1980). "Time-of-Day Corrections to Aircraft Noise Metrics". NASA CP 2135.
158. Taylor, S. M. Hall, F. L., and Bernie, S. E. (1980). "Effect of Background Levels on Community Response to Aircraft Noise". J. S. Vib. 71(2).
159. Anon. (1978). "A Guide to the Measurement and Prediction of the Equivalent Continuous Sound Level". Noise Advisory Council, HMSO, London.
160. Anon. (1977). "The Economic Impact of Noise". U.S. National Bureau of Standards.
161. Anon. (1980). "Environmental Health Criteria 12 (1980) Noise". World Health Organisation (WHO), Geneva.
162. Anon. (1980). "La Gêne Causée Par le Bruit Autours des Aeroports". Translation available as NASA TM-75784.
163. Bastenier, H., Klosterkoetter, W., and Large, J. B. (1975). "Damage and Annoyance Caused by Noise". Commission of the European Communities: EUR.5398e.
164. Emperly, J. A. (1987). Letter to FAA Rules Docket AGC-204 No. 25206 in response to FAA ANPRM 87-2: "Noise and Emission Standards for Aircraft Powered by Advanced Turboprop (Propfan) Engines".
165. Anon. (1975). "Concorde Supersonic Transport Aircraft: Final Environmental Impact Statement". DOT.
166. Ollerhead, J. B. (1973). "Scaling Aircraft Noise Perception". J. S. Vib. 26(3):361–88.
167. MAN – Acoustics and Noise. (1975). "Noise Certification Criteria and Implementation Considerations for V/STOL". FAA RD-75-190.
168. McCurdy, D. A., and Powell, C. A. (1984). "Annoyance Caused by Propeller Airplane Flyover Noise". NASA TP 2356.
169. McCurdy, D. A., and Powell, C. A. (1984). "Quantification of Advanced Turboprop Flyover Noise Annoyance". AIAA 84-2293.
170. McCurdy, D. A., Leatherwood, J. D., and Shepherd, K. P. (1986). "Advanced Turboprop Aircraft Noise Annoyance: A Review of Recent NASA Research". AIAA-86-1959.
171. McCurdy, D. A. (1988). "Annoyance Caused by Advanced Turboprop Aircraft Flyover Noise". NASA TP-2782.
172. Ollerhead, J. B. (1982). "Laboratory Studies of Scales for Measuring Helicopter Noise". NASA CR 3610.
173. Molino, J. A. (1982). "Should Helicopter Noise Be Measured Differently from Other Aircraft Noise?". NASA CR 3609.

General aeroacoustics and noise

200. Rayleigh, Lord. (1896). "The Theory of Sound". 2d ed., chap. 21. Macmillan. (Republished by Dover, 1945.)
201. Lighthill, M. J. (1954). "On Sound Generated Aerodynamically. General Theory (I) and Turbulence as a Source of Sound (II)". Proc. Roy. Soc. A221:564–87, A222:1–34.
202. Frey, A. R., and Kinsler, L. E. (1982). "Fundamentals of Acoustics". 3d ed. Wiley, Chichester.
203. Dowling, A. P., and Ffowcs-Williams, J. E. (1983). "Sound and Sources of Sound". Ellis Horwood Ltd., Chichester.
204. Pierce, A. D. (1986). "Acoustics; An introduction to its Physical Principles". McGraw-Hill, New York.
205. Goldstein, M. E. (1976). "Aeroacoustics". McGraw-Hill, New York.
206. Morse, P. M., and Ingard, K. V. (1968). "Theoretical Acoustics". McGraw-Hill, New York.

334   References

207. Goldstein, M. E. (1979). "General Aspects of the Theory of Aerodynamic Sound". Am. Phys. Soc. Fluid Symposium Paper AC2.
208. Cargill, A. M. (1983). "Sound Propagation through Fluctuating Flows – Its Significance in Aeroacoustics". AIAA 83-0697.
209. Kempton, A. J. (1977). "Efficient Sources of Aerodynamic Sound". Ph.D. thesis, Cambridge University.
210. White, R. G., and Walker, J. G. (Ed.). (1982). "Noise and Vibration". Ellis Horwood Ltd., Chichester.
211. Nelson, P. M. (Ed.). (1987). "Transportation Noise Reference Book". Butterworth, London.
212. Ffowcs Williams, J. E., and Hawkings, D. L. (1969). "Sound Generation by Turbulence and Surface in Arbitrary Motion". Phil. Trans. Roy. Soc. A264:321–42.
213. Dean, L. W. (1971). "Broadband Noise Generation by Aerofoils in Turbulent Flow". AIAA 71-587.
214. Muller, E. A. (Ed.). (1979). "Mechanisms of Sound Generation in Flows". Symp. Proc. Gottingen., Springer Verlag, Berlin.
215. Howe, M. S. (1975). "The Generation of Sound by Aerodynamic Sources in an Inhomogeneous Steady Flow". J. Fluid Mech. 67:597-610.
216. Cargill, A. M. (1983). "Sound Propagation through Fluctuating Flow – Its Significance in Aeroacoustics". AIAA 83-0697.
217. Kempton A. J. (1977). "Acoustic Scattering by Density Gradients". J. Fluid Mech. 83(3).
218. Dowling, A. P., Ffowcs Williams, J. E., and Goldstein, M. E. (1978). "Sound Production in a Moving Stream". Phil. Trans. Roy. Soc. (London) A288:321–49.
219. Powell, A. (1964). "Theory of Vortex Sound". J. Ac. Soc. Am. 36:177–95.
220. Campos, L. M. B. C. (1978). "The Spectral Broadening of Sound by Turbulent Shear Layers. Parts 1 and 2". J. Fluid Mech. 89:723–83.
221. Anon. (1975). "Acoustics". Progress in Aeronautics, vols. 37 and 38. American Institute of Aeronautics and Astronautics.
222. Evans, L. B., and Bazley, E. N. (1956). "The Absorption of Sound in Air at Audio Frequencies". Acoustica 6:238.
223. Harris, C. M. (1966). "Absorption of Sound in Air versus Humidity and Temperature". J. Ac. Soc. Am. 40:148–62.
224. Evans, L. B., Bass, H. E., and Sutherland, L. C. (1972). "Atmospheric Absorption of Sound: Theoretical Predictions". J. Ac. Soc. Am. 51:1565–75.
225. Anon. (1978). "Method for the Calculation of Absorption of Sound by the Atmosphere". American National Bureau of Standards, Washington, D.C., ANSI S126.
226. Piercy, J. E., Embleton, T. F. W., and Sutherland, L. C. (1977). "Review of Noise Propagation in the Atmosphere". J. Ac. Soc. Am. 61:1403–18.
227. Bazley, E. N. (1975). "Sound Absorption in Air up to 100 kHz". National Physical Laboratory, Eng. Report Ac. 74.
228. Sutherland, L. C., and Bass, H. E. (1979). "Influence of Atmospheric Absorption on the Propagation of Bands of Noise". J. Ac. Soc. Am. 66(3).
229. Anon. "Standard Values of Atmospheric Absorption as a Function of Temperature and Humidity for Use in Evaluating Aircraft Flyover Noise". SAE ARP 866A.
230. Anon. "Engineering Sciences Data: Noise Sub-series, Vol. 3: Sound Propagation". Engineering Sciences Data Unit (ESDU) International Ltd., London.
231. Howell, G. P., and Morphey, C. L. (1983). "Effects of Non-linear Propagation on Long-Range Noise Attenuation". AIAA 83-0700.
232. Sutherland, L. C. (1972). "Sound Propagation over Open Terrain from a Source Near the Ground". 84th Acoustical Soc. Am. Meeting.

233. Parkin, P. H., and Scholes, W. E. (1964). "The Horizontal Propagation of Sound from a Jet Engine Close to the Ground at Radlett". J. S. Vib. 1(1):1–13.
234. Parkin. P. H., and Scholes, W. E. (1965). "The Horizontal Propagation of Sound from a Jet Engine Close to the Ground at Hatfield". J. S. Vib. 2(4):353–74.
235. Pao, S. P., Wenzel, A. R., and Oncley, P. B. (1978). "Prediction of Ground Effects on Aircraft Noise". NASA TP-1104.
240. Davies, P. O. A. L. (1973). "Structure of Turbulence". J. S. Vib. 28:513–26.
241. Davies, P. O. A. L., and Yule, A. J. (1975). "Coherent Structures in Turbulence". J. Fluid Mech. 69:513-37.
242. Wille, R. (1963). "Growth of Velocity Fluctuations Leading to Turbulence in a Free Shear Layer". Hermann Fottinger Institut, Berlin. AFOSR Technical Report.
243. Anon. (1973). "Noise Mechanisms". AGARD Conference Preprint No. 13.
244. Anon. (1976). "Aircraft Noise Generation, Emission and Reduction". AGARD Lecture Series No. 77.
245. Anon. (1976). "Aerodynamic Noise". AGARD Lecture Series No. 80.

Jet noise

300. Lighthill, M. J. (1953). "On the Energy Scattered from the Interaction of Turbulence with Shock or Sound Waves". Proc. Cambridge Phil. Soc. 49:531–55.
301. Powell, A. (1953). "On the Mechanisms of Choked Jet Noise". Proc. Phys. Soc. 866:1139–50.
302. Ribner, H. S, (1955). "Shock Turbulence Interaction and the Generation of Sound". NASA Report 1233.
303. Hammit, A. G. (1961). "The Oscillations and Noise of an Overpressure Sonic Jet". Journal of Aerospace Sciences 28:673–80.
304. Hay, J. A., and Rose, E. G. (1970). "In-Flight Shock Cell Noise". J. S. Vib. 11:411–20.
305. Hardin, J. C. (1973). "Analysis of Noise Produced by an Orderly Structure of Turbulent Jets". NASA TN D-7242.
306. Fisher, M. J., and Harper-Bourne, M. (1973). "The Noise from Shock Waves in Supersonic Jets". Advisory Group for Aerospace Research and Development (AGARD) C. P. 131, "Noise Mechanisms".
307. Tanna, H. K. (1977). "An Experimental Study of Jet Noise, Part 2: Shock Associated Noise". J. S. Vib. 51:429–44.
308. Simcox, C. D. (1971). "Effects of Temperature and Shock Structure on Choked Jet Noise Characteristics". AIAA 71-582.
309. Seiner, J. M., and Norum, T. D. (1980). "Aerodynamic Aspects of Shock Containing Jet Plumes". AIAA-80-0965.
310. Ffowcs-Williams, J. E., Simpson, J., and Virchis, V. J. (1975). "Crackle – An Annoying Component of Jet Noise". J. Fluid Mech. 71.
311. Hansen, R. G. (1952). "The Noise Field of a Turbo Jet Engine". J. Ac. Soc. Am. 24(2).
312. Richards, E. J. (1953). "Research on Aerodynamic Noise from Jets and Associated Problems". RAeS Journal 57(509).
313. Hubbard, H. H., and Lassiter, L. W. (1953). "Experimental Studies of Jet Noise". J. Ac. Soc. Am. 25(3).
314. Anon. (Symposium) (1954). "Aeronautical Acoustics – In Particular, Jet Noise". RAeS Journal 58(520).
315. Greatrex, F. B. (1955). "Jet Noise". 5th International Aeroacoustics Conference.
316. Davies, P. O. A. L., Fisher, M. J., and Barrett, M. J. (1963). "Turbulence in the Mixing Region of a Round Jet". J. Fluid Mech. 15:337–67.
17. Lau, J. C., and Fisher, M. J. (1975). "The Vortex Sheet Structure of 'Turbulent' Jets: Part 1". J. Fluid Mech. 67:299–377.

318. Yule, A. J. (1978). "Large Scale Structure in the Mixing Layer of a Round Jet". J. Fluid Mech. 89:413–32.

319. Michalke, A. (1971). "Instabilitat eines Kompressiblen Runden Freistrahls". Z. Flugwiss 19:319–28.

320. von Glahn, U. H., Groesbeck, D., and Goodykoontz, J. (1973). "Velocity Decay and Acoustic Characteristics of Various Nozzle Geometries in Forward Flight". AIAA 73-629.

321. Moore, C. J., and Brierly, D. H. (1979). "Shear Layer Instability Noise Produced by Various Jet Nozzle Configurations". Proc. of Symposium on Mechanics of Sound Generated in Flows, Cottingen, Springer-Varlag, Berlin.

322 Moore, C. J. (1977). "The Role of Shear-Layer Instability Waves in Jet Exhaust Noise". J. Fluid Mech. 80(2).

323. Ffowcs-Williams, J. E. (1963). "Noise from Turbulence Convected at High Speed" Trans. Roy. Soc. A225:469–503.

324. Tanna, H. K., Dean, P. D., and Burrin, R. H. (1976). "The Generation of Radiation of Supersonic Jet Noise". U.S. Air Force Propulsion Laboratory Technical Report AFAPL-TR-76-65.

325. Tester, B. J., Morris, P. J., Lau, J. C., and Tanna, H. K. (1978). "The Generation, Radiation and Prediction of Supersonic Jet Noise". U.S. Air Force Aeropropulsion Laboratory Technical Report No. AFAPL-TR-78-85, Vol. 1.

326. Bushell, K. W. (1976). "Gas Turbine Engine Exhaust Noise". AGARD Lecture Series 80 – "Aerodynamic Noise".

327. Tanna, H. K. (1977). "Experimental Study of Jet Noise, Part 1: Turbulent Mixing Noise". J. S. Vib. 50:405–28.

328. Hoch, R. G., Duponchel, J. P., Cocking, B. J., and Bryce, W. D. (1973). "Studies of the Influence of Density on Jet Noise". J. S. Vib. 28(4):649–68.

329. Cocking, B. J. (1974). "The Effect of Temperature on Subsonic Jet Noise". National Gas Turbine Establishment (NGTE), Report No. R331.

330. Szewczyk, V., Morfey, C. M., and Tester, B. J. (1977). "New Scaling Laws for Hot and Cold Jet Mixing Noise Based on a High Frequency Model". AIAA 77-1287.

331. Morfey, C. L., Szewczyk, V. M., and Tester, B. J. (1978). "New Scaling Laws for Hot and Cold Jet Mixing Noise Based on a Geometric Acoustics Model". J. S. Vib., 61:255–92.

340. Hoch, R. G., and Berthelot, M. (1986). "Use of the Bertain Aerotrain for the Investigation of Flight Effects on Aircraft Engine Exhaust Noise". AIAA 76-534.

341. Drevet, P., Duponchel, J. P., and Jacques, J. R. (1976). "Effect of Flight on the Noise from a Convergent Nozzle as Observed on the Bertain Aerotrain". AIAA 76-557.

342. Smith, W. (1974). "The Use of a Rotating Arm Facility to Study Flight Effects on Jet Noise". Proc. Int. Symposium on Air Breathing Engines, Sheffield, Eng.

343. Cocking, B. J., and Bryce, W. D. (1975). "Subsonic Jet Noise in Flight Based on Some Recent Wind-Tunnel Tests". AIAA 75-462.

344. Cocking, B. J. (1976). "The Effect of Flight on Subsonic Jet Noise". AIAA 76-555.

345. Bryce, W. D., and Pinker, R. A. (1977). "The Noise from Unheated Supersonic Jets in Simulated Flight". AIAA 77-1327.

346. Way, D. J., and Francis, E. M. (1977). "The Simulation of Flight Effects on Jet Noise Using Co-Flowing Air Streams". AIAA 77-1305.

347. Ahuja, K. K., Tanna, H. K., and Tester, B. J. (1979). "Effects of Simulated Forward Flight on Jet Noise, Shock Noise and Internal Noise". AIAA 79-0615.

348. Blankenship, G. L., et al. (1977). "Effects of Forward Motion on Engine Noise". NASA CR-134954, October.

349. Low, J. K. (1977). "Effects of Forward Motion on Jet and Core Noise". AIAA 77-1330.

350. Bushell, K. W. (1975). "Measurement and Prediction of Jet Noise in Flight" AIAA 75-461.
351. Ahuja, K. K., Tester, B. J., and Tanna, H. K. (1978). "The Free Jet as a Simulator of Forward Velocity Effects on Jet Noise". NASA CR-3056.
352. Ahuja, K. K., Tester, B. J., and Tanna, H. K. (1981). "An Experimental Study of Transmission, Reflection and Scattering of Sound in a Free Jet Flight Simulation Facility and Comparison with Theory". J. S. Vib. 75:51–85.
353. Burrin, R. H., Ahuja, K. K., and Salikuddin, M. (1987). "High Speed Effects on Noise Propagation". AIAA-87-0013.
360. Goodykoontz, J. H. (1979). "Experimental Study of Coaxial Nozzle Exhaust Noise". AIAA 79-0631.
361. Szewczyk, V. M. (1979). "Co-axial Jet Noise in Flight". AIAA 79-0636.
362. Stevens, R. C. K., Bryce, W. D., and Szewczyk, V. M. (1983). "Model and Fullscale Studies of the Exhaust Noise from a Bypass Engine in Flight". AIAA 83-0751.
363. Cargill, A. M., and Duponchel, J. P. (1977). "The Noise Characteristics of Inverted Velocity Profile Co-annular Jets". AIAA 77-1263.
364. Packman, A. B., and Kozlowski, H. (1976). "Jet Noise Characteristics of Unsuppressed Duct Burning Turbofan Exhaust System". AIAA 76-149.
365. Pao, S. P. (1979). "A Correlation of Mixing Noise for Coannular Jets with an Inverted Velocity Profile". NASA TP-1301.
366. Strange, P. J. R., Podmore, G., Fisher, M. J., and Tester, B. J. (1984). "Coaxial Jet Noise Source Distributions". AIAA 84-2361.
370. Partharasathy, S. P., Cuffel, R., and Massier, P. F. (1978)."Influence of Internally Generated Pure Tones on the Broadband Noise Radiated from a Jet". AIAA 16(5):538–40.
371. Crow, S. C. (1972). "Acoustic Gain of a Turbulent Jet". Am. Phys. Soc. Paper 1E6.
372. Bechert, D., and Pfizenmaier, E. (1976). "On the Amplification of Broadband Jet Noise by a Pure Tone Excitation". AIAA 76-489.
373. Deneuville, P., and Jacques, J. R. (1977). "Jet Noise Amplification. A Practically Important Problem". AIAA 77-1368.
374. Ahuja, K. K. (1972). "An Experimental Study of Subsonic Jet Noise with Particular Reference to the Effects of Upstream Disturbances". M. Phil. thesis, University of London.
375. Way, D. T., and Turner, B. A. (1982). "Model Test Demonstrating Under-wing Installation Effects on Engine Exhaust Noise". AIAA 80-1048.
376. Southern, I. S. (1980). "Exhaust Noise in Flight: The Role of Acoustic Installation Effects". AIAA 80-1045.
377. Bashforth, S. (1981). "The Effect of Flight and the Presence of an Airframe on Exhaust Noise". AIAA 81-2029.
380. Lush, P. A. (1971). "Measurement of Subsonic Jet Noise and Comparison with Theory". J. Fluid Mech. 46(3):477–500.
381. Mani, R. (1972). "A Moving Source Problem Relevant to Jet Noise". J. S. Vib. 25: 337–47.
382. Lilley, G. M., Morris, P. J., and Tester, B. J. (1973). "On the Theory of Jet Noise and Its Applications". AIAA 73-987.
383. Ahuja, K. K., and Bushell, K. W. (1973). "An Experimental Study of Subsonic Jet Noise and Comparison with Theory". J. S. Vib. 30(3):317–41.
384. Ahuja, K. K. (1973). "Correlation and Prediction of Jet Noise". J. S. Vib. 29(2).
385. Davies, P. O. A. L., Hardin, J. C., Edwards, A. V. J., and Mason, J. P. (1973). "A Potential Flow Model for the Calculation of Jet Noise". AIAA 75-441.
386. Tester, B. J., and Morphey, C. L. (1975). "Developments in Jet Noise Modelling – Theoretical Predictions and Comparisons with Measured Data". AIAA 75-447.

387. Kempton, A. J. (1977). "Acoustic Scattering by Density Gradients." J. Fluid Mech. 83(3).

388. Ffowcs Williams, J. E., and Kempton, A. J. (1978). "The Noise from the Large-Scale Structure of a Jet". J. Fluid Mech. 84(4).

389. Tester, B. J., and Szewczyk, V. M. (1979). "Jet Mixing Noise – Comparison of Measurement and Theory". AIAA 79-0570.

390. Cargill, A. M. (1982). "The Radiation of High-Frequency Sound from a Jet Pipe". AIAA 80-0970 and J. S. Vib. 83(3).

391. Cargill, A. M. (1982). "Low-Frequency Sound Radiation due to the Interaction of Unsteady Flow with a Jet Pipe". J. Fluid Mech. 121.

392. Lu, H. Y. (1986). "An Emprical Model for Prediction of Coaxial Jet Noise in Ambient Flow". AIAA 86-1912.

393. Ahuja, K. K., Tester, B. J., and Tanna, H. K. (1987). "Calculation of Far Field Jet Noise Spectra from Near Field Measurements with True Source Location". J. S. Vib. 116(3):415–26.

Turbomachinery noise

400. Kemp, N. H., and Sears, W. R. (1953). "Aerodynamic Interference between Moving Blade Rows". Journal of the Aeronautical Sciences 20(9):585–97.

401. Kemp, N. H., and Sears, W. R. (1955). "The Unsteady Forces due to Viscous Wakes in Turbomachines". Journal of the Aeronautical Sciences 22(7).

402. Sharland, I. J. (1962). "Sources of Noise in Axial Flow Fans". J. S. Vib. 1(3):302–22.

403. Tyler, J. M., and Sofrin, T. G. (1962). "Axial Flow Compressor Noise Studies". SAE Transactions 70:309.

404. Hetherington, R. (February 1963). "Compressor Noise Generated by Fluctuating Lift Resulting from Rotor-Stator Interaction". AIAA Journal 1:473–74.

405. Bragg, S. L., and Bridge, R. (1964). "Noise from Turbojet Compressors". RAeS Journal 68:1–10.

406. Smith, M. J. T., and House, M. E. (1967). "Internally Generated Noise from Gas Turbine Engines – Measurement and Prediction". ASME Journal of Engineering for Power 89:177–90.

407. Lowson, M. V. (1970). "Theoretical Analysis of Compressor Noise". J. Ac. Soc. Am. 47(1):371–85.

408. Morfey, C. L. (1972). "The Acoustics of Axial Flow Machines". J. S. Vib. 22(4):445–66.

409. Hanson, D. B. (1974). "Spectrum of Rotor Noise Caused by Inlet Guide Vane Wakes". J. Ac. Soc. Am. 55(6):1247–51.

410. Kaji, D., and Okazaki, T. (1970). "Propagation of Sound Waves Through a Blade Row". J. S. Vib. 11(3):355–75.

411. Philpot, M. G. (1975). "The Role of Rotor Blockage in the Propagation of Fan Noise Interaction Tones". AIAA 75-447.

412. Cumpsty, N. A. (1977). "A Critical Review of Turbomachinery Noise". J. Fluids Eng., Trans. ASME 99, ser. 1 (2):278–93.

420. Barry, B., and Moore, C. J. (1971). "Subsonic Fan Noise". J. S. Vib. 17(2).

421. Lowrie, B. W. (1976). "Fan Noise". AGARD Lecture Series No. 80.

422. Morfey, C. L. (1973). "Rotating Blades and Aerodynamic Sound". J. S. Vib. 28(3):587–617.

423. Mather, J. S. B., Savidge, J., and Fisher, M. J. (1971). "New Observations on Tone Generation in Fans". J. S. Vib. 16.

424. Lowson, M. V., Whatmore, A. R., and Whitfield, C. E. (1973). "Source Mechanisms for Rotor Noise Radiation". NASA CR-2077.

425. Hanson, D. B. (1975). "Study of Noise Sources in a Subsonic Fan Using Measured Blade Pressures and Acoustic Theory". NASA CR-2574.
426. Mugridge, B. D., and Morfey, C. L. (1975). "Sources of Noise in Axial Flow Fans". J. Ac. Soc. Am. 5(5):Pt.I.
430. Moore, C. J. (1972). "In-duct Investigation of Subsonic Fan Rotor Alone Noise". J. Ac. Soc. Amer. 55(1).
431. Metzger, F. B. (1973). "Analytical Parametric Investigations of Low Pressure Ratio Fans". NASA CR-2188.
432. Magliozzi, B. (1973). "Noise and Wake Structure Measurements in a Subsonic Tipspeed Fan". NASA CR-2323.
433. Kaji, S., and Okazaki, T. (1970). "Generation of Sound by Rotor-Stator Interaction". J. S. Vib. 13(3):281–307.
434. Cumpsty, N. A. (1972). "Tone Noise from Rotor/Stator Interactions in High Speed Fans". J. S. Vib. 24(3):393–409.
435. Schwaller, P. J. G., Parry, A. B., Oliver, M. J., and Eccleston, A. (1984). "Farfield Measurement and Mode Analysis of the Effect of Vane/Blade Ratio on Fan Noise". AIAA 84-2280.
436. Moore, C. J. (1972). "Analysis of Fan Noise In-Ducts". Brit. Acoustical Soc. 2(1).
437. Moore, C. J. (1980). "The Asymmetric Radiation Patterns of Interacting Duct Modes". Acoustica 47(1).
438. Osborne, C. (1973). "Compressible Unsteady Interactions between Blade Rows". AIAA Journal 11:340–6.
439. Ffowcs Williams, J. E., and Hawkings, D. L. (1969). "Theory Relating to the Noise of Rotating Machinery". J. S. Vib. 10(1):10–21.
440. Mani, R. (1970). "Discrete Frequency Noise Generation from an Axial Flow Fan Blade Row". Journal of Basic Engineering Trans. ASME ser. D 92:37–43.
441. Lipstein, N. J., and Mani, R. (1972). "Experimental Investigation of Discrete Frequency Noise Generated by Unsteady Blade Forces". J. Basic Engineering, Trans. ASME ser. D 92:155–64.
442. Hanson, B. (1973). "Unified Analysis of Fan Stator Noise". J. Ac. Soc. Am. 54(6):1571–91.
443. Morfey, C. L. (1970). "Broadband Sound Radiated from Subsonic Rotors". NASA SP-304, Pt. II.
444. Ginder, R. B., and Newby, D. R. (1976). "A Study of Factors Affecting the Broadband Noise of High Speed Fans". AIAA 76-567.
445. Mugridge, B. D. (1971). "Acoustic Radiation from Aerofoils with Turbulent Boundary Layers". J. S. Vib. 16(4).
446. Ginder, R. B., and Newby, D. R. (1977). "An Improved Correlation for the Broadband Noise of High Speed Fans". J. of Aircraft 14:844–9.
450. Kaji, S. (1975). "Noncompact Source Effect on the Prediction of Tone Noise from a Fan Rotor". AIAA 75-446.
451. Morfey, C. L., and Fisher, M. J. (1970). "Shock-Wave Radiation from a Supersonic Ducted Rotor". Aeronautical Journal 74:579–85.
452. Sofrin, T. G., and Pickett, G. F. (1970). "Multiple Pure Tone Noise Generated by Fans at Supersonic Tip Speeds". NASA SP-304. Pt. II.
453. Hawkings, D. (1971). "Multiple Tone Generation by Transonic Compressors". J. S. Vib. 17(2):241–50.
454. Kurosaka, M. (1971). "A Note on Multiple Pure Tone Noise". J. S. Vib. 19(4):453–62.
455. Stratford, B. S., and Newby, D. R. (1977). "A New Look at the Generation of Buzz-Saw Noise". AIAA 77-1343.
56. Cargill, A. M. (1983). "Shock Waves ahead of a Fan with Non-Uniform Blades". AIAA Journal 21.

460. Mani, R. (1973). "Noise due to Interaction of Inlet Turbulence with Isolated Stators and Rotors". J. S. Vib. 17(2):251–60.
461. Cumpsty, N. A., and Lowrie, B. W. (1974). "The Cause of Tone Generation by Aero-Engine Fans at High Subsonic Speeds and the Effect of Forward Speed." ASME J. Eng. for Power 96(3):228–34.
462. Hanson, D. B.(1974). "Spectrum of Rotor Noise Caused by Atmospheric Turbulence". J. Ac. Soc. Am. 56(1):110–26.
463. Pickett, G. F. (1974). "Effects on Non-uniform Inflow on Fan Noise". Spring Meeting of Ac. Soc. Am.
464. Feiler, C. E., and Merriman, J. E. (1974). "Effects of Forward Velocity and Acoustic Treatment on Inlet Fan Noise". AIAA 74-946.
465. Hanson, D. B. (1975). "Measurements of Static Inlet Turbulence". AIAA 75-467.
466. Grosche, F. R., and Stiewitt, H. (1977). "Investigation of Rotor Noise Source Mechanisms with Forward Speed". AIAA Journal 17(12).
467. Roundhill, J. P., and Schaut, L. A. (1975). "Model and Full Scale Test Results Relating to Fan Noise In-flight Effects". AIAA 75-465.
468. Lowrie, B. W. (1975). "Simulation of Flight Effects on Aero-Engine Fan Noise". AIAA 75-463.
469. Cocking, B. J., and Ginder, R. B. (1977). "The Effect of an Inlet Flow Conditioner on Fan Distortion Tones". AIAA 77-1324.
480. Kempton, A. J. (1979). "Ray-Theory Approach for High-Frequency Engine-Intake Noise". Proc. of Symposium on Mechanics of Sound Generated in Flows, Gottingen. Springer-Verlag, Berlin.
481. Kempton, A. J. (1980). "Ray-Theory to Predict the Propagation of Broadband Fan-Noise". AIAA 80-0968.
482. Kempton, A. J., and Smith, M. G. (1981). "Ray-Theory Predictions of the Sound Radiated from Realistic Engine Intakes". AIAA 81-1982.
490. Smith, M. J. T. (1968). "The Problem of Turbine Noise in the Civil Gas Turbine Engine". ICAS 68-35.
491. Smith, M. J. T., and Bushell, K. W. (1969). "Turbine Noise – Its Significance in the Civil Aircraft Noise Problem". ASME Paper 69-WA/ST-12.
492. Kazin, S. B., Matta, R. K., et al. (1974). "Core-Engine Noise Control Program: Volume I – Identification of Component Noise Sources". FAA-RD-74-125.
493. Mathews, D. C., and Peracchio, A. A. (1974). "Progress in Core Engine and Turbine Noise Technology". AIAA 74-948.
494. Kazin, S. B., and Matta, R. K. (1975). "Turbine Noise Generation, Reduction and Prediction". AIAA 75-449.
495. Fletcher, J. S., and Smith, P. H. (1975). "The Noise Behaviour of Aero Engine Turbine Tones". AIAA 75-466.
496. Pickett, G. F. (1974). "Turbine Noise Due to Turbulence and Temperature Fluctuations". Paper to 8th International Congress on Acoustics, London.
497. Cumpsty, N. S., and Marble, F. E. (1977). "The Interaction of Entropy Fluctuations with Turbine Blade Rows: A Mechanism of Turbojet Engine Noise". Proc. Roy. Soc. London, ser. A. 357:323–44.

Duct acoustics and absorbent liners

500. Vaidya, P. G., and Dean, P. D. (1977). "The State of the Art of Duct Acoustics". AIAA 77-1279.
501. Morfey, C. L. (1964). "Rotating Pressure Patterns in Ducts; Their Generation and Transmission". J. S. Vib. 1:60–87.
502. Morfey, C. L. (1969). "A Note on the Radiation Efficiency of Acoustic Duct Modes". J. S. Vib. 9:367–72.

503. Morfey, C. L. (1971). "Sound Generation and Transmission in Ducts with Flow". J S. Vib. 14:37–55.
504. Mungur, P., and Gladwell, G. M. L. (1969). "Acoustic Wave Propagation in a Sheared Fluid Contained in a Duct". J. S. Vib. 9:28–48.
505. Savkar, S. D. (1975). "Radiation of Cylindrical Duct Acoustic Modes with Flow Mismatch." J. S. Vib. 42(3):363–86.
506. Nayfeh, A. H., Sun, J., and Telionis, D. P. (1974). "Effects of Transverse Velocity and Temperature Gradients on Sound Attenuation in Two-dimensional Ducts". J. S. Vib. 34(4):505–17.
507. Nayfeh, A. H., and Tsai, M. S. (1974). "Non-linear Acoustic Propagation in Two-dimensional Ducts". J. Ac. Soc. Am. 55(6):1100–72.
508. Kapur, A., and Mungur, P. (1972). "On the Propagation of Sound in a Rectangular Duct with Gradients of Mean Flow and Temperature in Both Transverse Directions". J. S. Vib. 23:401–4.
509. Cummings, A. J. (1974). "Sound Transmission in Curved Duct Bends". J. S. Vib. 35:451–77.
510. Cummings, A. J. (1975). "Sound Transmission in a Folded Annular Duct". J. S. Vib. 41:375–9.
511. Lansing, D. L., Drischler, J. A., and Pusey, C. G. (1970). "Radiation of Sound from an Unflanged Circular Duct with Flow", Paper to 79th Meeting of Acoustical Society of America.
512. Mungur, P., Plumblee, H. E., and Doak, P. E. (1974). "Analysis of Acoustic Radiation in a Jet Flow Environment". J. S. Vib. 36:21–52.
513. Davies, P. O. A. L., and Halliday, R. F. (1981). "Radiation of Sound by a Hot Exhaust". J. S. Vib. 76:591–4.
530. Scott, R. A. (1946). "The Propagation of Sound between Walls of Porous Material". Proc. Phys. Soc. Lond. 58:358–68.
531. Rice, E. J. (1968). "Attenuation of Sound in Soft Walled Ducts". NASA TMX-52442.
532. Bokor, A. (1969). "Attenuation of Sound in Lined Ducts".J. S. Vib. 10(3):390–403.
533. Snow, D. J., and Lowson, M. V. (1975). "Attenuation of Spiral Modes in a Circular and Annular Lined Duct". J. S. Vib. 25(3).
534. McCormick, M. A. (1975). "The Attenuation of Sound in Lined Rectangular Ducts Containing Uniform Flow". J. S. Vib. 39(1).
535. Mungur, P., and Plumblee, H. E., Jr. (1969). "Propagation and Attenuation of Sound in a Soft-walled Annular Duct Containing a Sheared Flow. Basic Aerodynamic Noise Research". NASA, SP-207:305–27.
536. Nayfeh, A. H., Sun., J., and Telionis, D. P. (1974). "Effect of Bulk-reacting Liners on Wave Propagation in Ducts". AIAA Journal, 12(6):838–43.
537. Atvars, J., and Mangiarotty, R. A. (1970). "Parametric Studies of the Acoustic Behaviour of Duct Lining Materials". J. Ac. Soc. Am. 48(3): Pt. 3, 815–25.
538. Kurtze, U. J., and Ver, I. L. (1972). "Sound Attenuation in Ducts Lined with Non-isotropic Material". J. S. Vib., 24(2):177–287.
539. Tester, B. J. (1972). Sound Attenuation in Lined Ducts Containing Subsonic Mean Flow". Ph.D. Thesis, University of Southampton.
540. Ko, S. H. (1972). "Sound Attenuation in Acoustically Lined Circular Ducts in the Presence of Uniform Flow and Shear Flow". J. S. Vib. 22:193–210.
541. Tack, D. M., and Lambert, R. F. (1965). "Influence of Shear Flow on Sound Attenuation in Lined Ducts". J. Ac. Soc. Am. 38(4).
542. Tester, B. J. (1973). "The Optimization of Modal Sound Attenuation in Ducts, in the Absence of Mean Flow". J. S. Vib. 27:477–513.
543. Kurtze, U. J. and Allen, C. M. (1971). "Influence of Flow and High Sound Level on the Attenuation in a Lined Duct". J. Ac. Soc. Am. 49(5):Pt. 2, 1943–7.

544. Candell, S. (1973). "Acoustic Radiation from the End of a 2-Dimensional Duct. Effects of Uniform Flow and Duct Lining". J. S. Vib. 28(1):1–13.
550. Plumblee, H. E., Dean, P. D., Wynne, G. A., and Burrin, R. H. (1973). "Sound Propagation in the Radiation from Acoustically Lined Flow Ducts; A Comparison of Experiment and Theory". NASA CR-2306.
551. Webber, C. J. (1980). "The Development of Acoustic Absorbers for Turbofan Engines". AGARD CP 42 "Aircraft Engine Noise and Sonic Boom".
552. Syed, A. A., and Bennett, S. C. (1978). "Comparison of Measured Broadband Noise Attenuation Spectra from Circular Flow Ducts and from Lined Engine Intakes with Predictions". J. S. Vib. 56(4).
553. Kempton, A. J. (1983). "Ray-Theory and Mode-Theory Predictions of Intake-Liner Performance – A Comparison with Engine Measurements". AIAA 83-0711.
554. Delaney, M. E., and Bazeley, E. N. (1969). "Acoustic Characteristics of Fibrous Absorbent Materials". NPL Aero Report Ac. 37.
555. Armstrong, D. L. (1971). "Acoustic Grazing Flow Impedance Using Waveguide Principles" NASA CR-120848.
556. Nayfeh, A. H., Kaiser, J. E., and Telionis, D. P. (1975). "The Acoustics of Aircraft Engine-Duct Systems". AIAA Journal 13:130–53.
577. Zorumski, W. E. (1973). "Acoustic Impedance of Curved Multilayered Duct Liners". NASA-TN-D-7277.

Propeller noise

600. Gutin, L. J. (1936). "On the Sound Field of a Rotating Propeller". Physikalische Zeitschrift det Sowjetunion 9(1). Translated as NACA Tech. Memo 1195.
601. Hicks, L., and Hubbard, H. H. (1947). "Comparision of Sound Emission from Two-Blade, Four-Blade, and Seven-Blade Propellers". NACA TN 1354.
602. Hubbard, H. H., and Lassiter, L. W. (1952). "Sound from a Two-Blade Propeller at Supersonic Tip Speeds". NACA Report 1079.
603. Garrick, I. E., and Watkins, C. E. (1954). "A Theoretical Study of the Effect of Forward Speed on the Free-Space Sound Pressure Field around Propellers". NACA Report 1198. (See also Watkins and Durling [1956], NACA TN 3809.)
604. Morfey, C. L. (1973) "Rotating Blades and Aerodynamic Sound". J. S. Vib. 28:587–17.
605. Heller, H., Kallergis, M., Ahlswede, M., and Dobrzynski, W. (1979). "Rotational and Vortex-Noise of Propellers in the 100 to 150 kW class". AIAA 79-1611.
606. Patterson, R. W., et al. (May 1973). "Vortex Noise of Isolated Airfoils," Journal of Aircraft 10(4):296–302.
607. Hanson, D. B. (1979). "The Influence of Propeller Design Parameters on Farfield Harmonic Noise in Forward Flight". AIAA 79-0609.
608. Williams, J. (Ed.) (1982). "Propeller Performance and Noise". Ven Karman Institute (Belgium) Lecture Series Publication 1982-08.
609. Anon. (1984). "Aerodynamics and Acoustics of Propellers". AGARD Conference Proceedings No. 366.
610. Magliozzi, B. (1977). "The Influence of Forward Flight on Propeller Noise". NASA CR-145105.
611. Trebble, W. J. G., Williams, K. J., and Donnelly, R. P. (1981). "Comparative Acoustic Wind Tunnel Measurements and Theoretical Correlation on Subsonic Aircraft Propellers: Full Scale and Model Scale". AIAA 81-2004.
612. Bonneau, H., Wilford, D. F., and Wood, L. (1984). "An Investigation of In-flight Near-Field Propeller Noise Generation and Transmission". AGARD Conference Proceedings No. 366.
613. Grosche, F. R., and Stiewitt, H. (1984). "Aeroacoustic Wind Tunnel Measurements on Propeller Noise". AGARD Conference Proceedings No. 366.

614. Trebble, W. G., Williams, J., and Donnelly, R. P. (1984). "Some Aeroacoustic Wind Tunnel Measurements, Theoretical Predictions and Flight Test Correlations on Subsonic Aircraft Propellers". AGARD Conference Proceedings No. 366.

615. Kallergis, M., and Neubauer, R. F. (1983). "Propeller Nearfield and Noise Radiation to the Farfield". Proc. InterNoise.

616. Bass, R. M. (1983). "An Historical Review of Propeller Developments". RAeS Aeron. Journal, Paper No. 1089.

617. Hanson, D. B. (1987). "Propeller and Propfan Aeroacoustics". AIAA 87-2655.

620. Magliozzi, B. (1984). "Advanced Turboprop Noise: A Historical Review". AIAA 84-2261.

621. Parzych, D. J., Magliozzi, B., and Metzger, F. B. (1987). "Prop-Fan/Turboprop Acoustic Terminology". SAE 871839.

623. Hawkings, D. L., and Lowson, M. V. (1975). "Noise of High Speed Rotors". AIAA 75-450.

624. Brooks, B. M., and Metzger, F. B. (1979). "Acoustic Test and Analysis of Three Advanced Turboprop Models". Final Report NASA Contract NAS3-20614.

625. Hanson, D. B. (1976). "Near Field Noise of High Speed Propellers in Forward Flight". AIAA 76-565.

626. Hawkings, D. L., and Lowson, M. V. (1974). "Theory of Open Supersonic Rotor Noise". J. S. Vib. 36(11):1–20.

627. Glegg, S., and Wills, C. R. (1979). "High Speed Rotor Thickness Noise". Proc. Inst. Ac. 20.GS.

628. Anon. (1984) "Aerodynamics and Acoustics of Propellers". AGARD Conference Proceedings No. 336.

629. Farassat, F., and Succi, G. P. (1980). "A Review of Propeller Discrete Frequency Technology with Emphasis on Two Current Mehods for Time-Domain Calculations". J. S. Vib. 71:399–419.

630. Hanson, D. B. (1983). "Compressible Helicoidal Surface Theory for Propeller Aerodynamics and Noise". AIAA Journal 21:881–89.

631. Farassat, F. (1984). "The Unified Acoustic and Aerodynamic Prediction of Advanced Propellers in the Time Domain". AIAA 84-2303.

632. Parry, A. B., and Crighton, D. G. (1986). "Theoretical Prediction of Single-Rotation Propeller Noise". AIAA 86-1891.

633. Watanabe, T. (1987). "Prediction of Counter-rotation Propeller Noise". AIAA 87-2658.

640. Bradley, A. J. (1986). "A Study of the Rotor/Rotor Interaction Tones from a Contra-Rotating Propeller-Driven Aircraft". AIAA 86-1894.

641. Hanson, D. B. (1984). "Noise of Counter-Rotation Propellers". AIAA 84-2305.

650. Gregorek, G. M., Hoffmann, M. J., and Newman, R. L. (1984). "Flight Measurements on the Effect of Angle of Attack on Propeller Acoustics". AIAA 84-2348.

651. Padula, S. L., and Block, P. J. W. (1985). "Predicted Changes in Advance Turboprop Noise with Shaft Angle of Attack." J. Propulsion 1(5):381–7.

652. Burrin, R. H., and Salikuddin, M. (1983), "Sources of Installed Propeller Noise". AIAA 83-0744.

653. Block, P. J. W. (1984). "Installation Noise of Model SR and CR Propellers". NASEA TM 8570.

654. Heidelberg. L. J., Rice, E. J., and Dahl, M. D. (1983). "Installation Effects on the Noise of a Model High Speed Propeller". AIAA 84-2346.

655. Wilford, D. F., and Bonneau, H. (1983). Flyover Noise Measurements for Turbo-Prop Aircraft". AIAA 83-0746.

680. Leverton, J. W. (1971). "The Sound of Rotorcraft". RAeS Aeron. Journ 75:385-97.

681. Leverton, J. W., Pollard, J. S., and Wills, C. R. (1975). "Main Rotor Wake/Tail Rotor Interaction". First European Rotorcraft Forum, Paper 25.

682. Schmitz, F. H., and Yu, Y. H. (1983). "Helicopter Impulsive Noise: Theoretical and Experimental Status". NASA TM 84390.
683. Splettstoesser, W. R., Schultz, K. J., Boxwell, D. A., and Schmitz, F. H. (1984). "Helicopter Model Rotor-Blade Vortex Interaction Noise: Scalability and Parametric Variations". Tenth European Rotorcraft Forum, Paper 18.
684. Schlegel, R. G., King, R. J., and Mull, H. R. (1966). "Helicopter Rotor Noise Generation and Propagation". U.S.A. AVLABS Tech. Rep. 66-4.
685. Lowson, M. V., and Ollerhead, J. B. (1969). "Studies of Helicopter Rotor Noise". U.S.A. AVLABS Report 68-80.
686. Lowson, M. V., and Ollerhead, J. B. (1969). "A Theoretical Study of Helicopter Rotor Noise". J. S. Vib. 9:197–222.
687. Lowson, M. V., Whatmore, A. R., and Whitfield, L. E. (1973). "Source Mechanisms for Rotor Noise Radiation". NASA CR 2077.
688. Tadghighi, H., and Cheeseman, I. C. (1983). "A Study of Helicopter Rotor Noise, with Special Reference to Tail Rotors, Using an Acoustic Windtunnel". Vertica 7:9–32.
689. Wright, S. E. (1969). "Sound Radiation from a Lifting Rotor Generated by Asymmetric Disk Loading". J. S. Vib. 9:223–40.
690. George, A. R., and Chou, S. T. (1983). "Comparison of Broadband Noise Mechanisms, Analyses, and Experiments on Helicopters, Propellers and Wind Turbines". AIAA Paper 83-0690.
691. Aravamundan, K. S., and Harris, W. L. (1979). "Low Frequency Broadband Noise Generated by a Model Rotor". J. Ac. Soc. Am. 66(2):522.
692. Aravamundan, K. S., Lee, A., and Harris, W. L. (1978). "A Simplified Mach Number Scaling Law for Helicopter Rotor Noise". J. S. Vib. 57(4):555–70.
693. Paterson, R. W., and Amiet, R. K. (1979). "Noise of a Model Helicopter Rotor due to Ingestion of Turbulence". NASA CR-3213.
694. Hubbard, H. H., and Maglieri, D. J. (1960). "Noise Characteristics of Helicopter Rotors at Tip Speeds up to 900 Feet per Second". J. Ac. Soc. Am. 32: 1105–7.
695. Anon. (1987). "Helicopter Noise Measurement Repeatability Programme". Final Report. FAA-EE-87-2.

Other noise sources

700. Gibson, J. S. (1974). "Recent Developments at the Ultimate Noise Barrier". ICAS 74-59.
701. Hardin, J. C. (1976). "Airframe Self-noise – Four Years of Research". NASA TMX-73908.
702. Heller, H. H., and Dobrzynski, W. M. (1978). "A Comprehensive Review of Airframe Noise Research". ICAS Paper No. GL-03.
703. Paterson, R. W., Vogt, P. G., Fink, M. R., and Munch, C. L. (1973). "Vortex Noise of Isolated Aerofoils". Journal of Aircraft, 10:296–302.
704. Revell, J. D. (1975). "Induced Drag Effect on Airframe Noise". AIAA 75-497.
705. Heller, H. H., and Dobrzynski, W. M. (1976). "Sound Radiation from Aircraft Wheel-Well/Landing Gear Configurations". AIAA 76-552.
706. Tam, C. K. W. (1974). "Discrete Tones of Isolated Airfoils". J. Ac. Soc. Am. 55(6):1173–7.
707. Fethney, P. (1975). "An Experimental Study of Airframe Self-noise". AIAA 75-511.
708. Fethney, P., and Jelly, A. H. (1980). "Airframe Self-noise Studies on the Lockheed L1011 TriStar Aircraft". AIAA 80-1061.
709. Hayden, R. E., Kadman, Y., Bliss, D. B., and Africk, S. A. (1975). "Diagnostic Calculations of Airframe-Radiated Noise". AIAA 75-485.

710. Fink, M. R., and Schinker, R. H. (1979). "Airframe Noise Component Interaction Studies". NASA CR-3110.
720. Witham, G. B. (1952). "The Flow Pattern of a Supersonic Projectile". Communication on Pure and Applied Mathematics, 5:301–48.
721. Warren, C. H. E. (1954). "Noise Associated with Supersonic Flight". JRAes 58.
722. Whitham, G. B. (1956). "On the Propagation of Weak Shock Waves". J. Fluid Mech. 1:Pt. 3.
723. Walkden, F. (1958). "The Shock Pattern of a Wing-Body Combination, Far from the Flight Path". Aero. Quart. 9:Pt. 2.
724. Struble, R. A., Stewart, C. E., Brown, E. A., and Ritten, A. (1957). "Theoretical Investigations of Sonic Boom Phenomena". WADC Tech. Rep. 57-412.
725. Hayes, W. D. (1971). "Sonic Boom". Annual Review of Fluid Mechanics 3:269–90.
726. Morris, J. (1960). "An Investigation of Lifting on the Intensity of Sonic Booms". JRAeS 64.
727. Ferri, A., Siclari, M., and Ting, L. (1973). "Sonic Boom Analysis for High Altitude Flight at High Mach Number". AIAA 73-1034.
728. Anon. (1972 and 1973). "ICAO Sonic Boom Committee". Reports of 1st and 2d Meetings, ICAO Docs 9011, 9064.
729. Nixon, C. W., and Hubbard, H. H. (1965). "Results of USAF-NASA-FAA Flight Program to Study Community Responses to Sonic Booms in the Greater St. Louis Area". NASA TN D-2705.
730. Borsky, P. N. (1972). "Sonic Boom Exposure Effect". J. S. Vib. 20:527.
731. Thackray, R. I., Touchstone, R. M., and Bailey, J. P. (1974). "A Comparison of the Startle Effects Resulting from Exposure to Two Levels of Simulated Sonic Boom". J. S. Vib. 33:379–89.
732. Rickley, E. J., and Pierce, A. D. (1980) "Detection and Assessment of Secondary Sonic Booms in New England". FAA-AEE-80-22.
733. Hershey, R. L., Kevala, R. J., and Burns, S. L. (1975). "Analysis of the Effect of Concorde Aircraft Noise on Historic Structures". FAA-RD-75-118.
734. Clarkson, B. L., and Mayes, W. H. (1972). "Sonic-boom-induced Building Structure Responses including Damage". J. Ac. Soc. Am. 51:Pt. 3, 742–57.
735. Anon. (1974). "Operational Aspects of the Sonic Boom". British Aircraft Corporation and Aerospatiale joint publication No. C606.
736. Borsky, P. N. (1965). "Community Reaction to Sonic Booms in the Oklahoma City Area". National Opinion Research Centre AMRL-TR-65-37.
737. Anon. (1967). "Sonic Boom: A Review of Current Knowledge and Developments". Boeing Document D6A10598-1.
740. Bragg, S. L. (1963). "Combustion Noise". Journal of the Institute of Fuel 36: 12–16.
741. Grande, E. (1973). "A Review of Core Engine Noise". AIAA 73-1026.
742. Strahle, W. C., and Shivashankara, B. N. (1976). "Combustion Generated Noise in Gas Turbine Combustors". Journal of Engineering for Power. Trans. ASME 98:242–6.
743. Crighton, D. G. (1975). "Mechanisms of Excess Jet Noise". Conference on Noise Mechanisms, AGARD CP-131.
744. Gerend, R. P., Kumasake, H. A., and Roundhill, J. P. (1973). "Core Engine Noise". AIAA 73-1027.
745. Strahle, W. C. (1973). "A Review of Combustion Generated Noise". AIAA 73-1023.
46. Arnold, J. S. (July 1972). "Generation of Combustion Noise". J. Ac. Soc. Am. 52(1): Pt. 1, 5–12.
47. Chiu, H. H., and Summerfield, M. (1974). "Theory of Combustion Noise". Acta Astronautica 1:967–84.

346 *References*

748. Hurle, I. R., Price, R. B., Sugden, T. M., and Thomas, A. (1968). "Sound Emission from Open Turbulent Premixed Flames". Proc. Roy. Soc. London ser. A, 303:409–27.
749. Shivashankara, B. N., Strahle, W. C., and Handley, J. C. (1973). "Combustion Noise Radiation by Open Turbulent Flames". AIAA 73-1025.
750. Strahle, W. C. (1971). "On Combustion Generated Noise". J. Fluid Mech. 48:Pt. 2, 399–414.
751. Cumpsty, N. A., and Marble, F. E. (1977). "Core Noise from Jet Engines". J. S. Vib. 54(3):297–309.
752. Strahle, W. C. (1975). "The Convergence of Theory and Experiment in Direct Combustion Generated Noise". AIAA 75-522.
753. Abdelhamid, A. N., Harrje, D. T. Plett, E. G., and Summerfield, M. (1973). "Noise Characteristics of Combustion Augmented High-Speed Jets". AIAA 73-189.
754. Pickett, G. F. (1975). "Core Engine Noise due to Temperature Fluctuations Convecting through Turbine Blade Rows". AIAA 75-528.
755. Kazin, S. B., and Emmerling, J. J. (1974). "Low Frequency Core Engine Noise". ASME 74-WA/Aero.2.
756. Chiu, H. H., Plett, E. G., and Summerfield, M. (1973). "Noise Generated by Ducted Combustion Systems". AIAA 73-1024.
757. Hoch, R. G., Thomas, P., and Weiss, E. (1975). "An Experimental Investigation of the Core Engine Noise of a Turbofan Engine". AIAA paper 75-526.
760. Bryce, W. D., and Stevens, R. C. K. (1975). "An Investigation of the Noise from a Scale Model of an Engine Exhaust System". AIAA 75-459.
761. Bryce, W. D. (1979). "Experiments Concerning the Anomalous Behaviour of Aero Engine Exhaust Noise in Flight". AIAA 79-0648.
780. Hehmann, H. W. (1973). "Self-Generated Noise of Duct Treatment". 86th Meeting, Acoustical Society of America.
781. Tsui, C. Y., and Flandro, G. A. (1977). "Self-induced Sound Generation by Flow over Perforated Duct Liners". J. S. Vib. 50:315–31.
782. Bauer, A. B., and Chapkis, R. L. (1977). "Noise Generated by Boundary Layer Interaction with Perforated Acoustic Liners". Journal of Aircraft 14:157–60.
790. Higginson, R. F., and Rennie, R. J. (1977). "Noise from Engine Thrust Reversal on Landing". National Physical Lab., Eng. Report Ac. 83.

Aircraft noise control

800. Harris, C. M. (1957). "Handbook of Noise Control". McGraw-Hill, New York.
801. Beranek, L. L. (1960). "Noise Reduction". McGraw Hill, New York.
802. Beranek, L. L. (Ed.) (1971). "Noise and Vibration Control". McGraw Hill, New York.
803. Richards, E. J., and Mead, D. J. (1968). "Noise and Acoustic Fatigue". John Wiley and Sons, New York.
804. Anon. (1981). "Assessment of Technological Progress Made in Reduction of Noise from Subsonic and Supersonic Jet Aeroplanes". ICAO Circular 157-AN/101.
805. Bragdon, C. R. (1974). "Environmental Noise Control Programs in the United States". J. S. Vib. 11(12).
806. Harris, A. S. (1980). "Designing for Noise Control at Air Carrier Airports: Runway Layout and Use". Noise Control Engineering.
807. Greatrex, F. B. (1965). "Reducing the Annoyance from Aircraft Noise". Discovery, May 1965.
808. Greatrex, F. B. (1966). "The Economics of Aircraft Noise Suppression". ICAS, 66-5.

809. Dawson, L. G., and Sills, T. D. (1972). "An End to Aircraft Noise?" JRAeS 76.
810. Smith, M. J. T. (1973). "Quieter Aero-Engines: Cause and Effect". British Acoustical Society 1972 Silver Medal Address.
811. Smith, M. J. T. (1982). "Has Anybody Noticed the Improving Aircraft Noise Situation?". RAeS Aerospace Journal, 9(6).
812. Smith, M. J. T. (1986). "Aircraft Noise Control". World Aerospace Profile.
820. Greatrex, F. B., and Brown, D. M. (1958) "Progress in Jet Engine Noise Reduction". ICAS, First Congress, Madrid.
821. Greatrex, F. B. (1960). "By-Pass Engine Noise". SAE National Aeronautic Meeting, New York.
822. Conrad, E. W., and Ciepluch, C. C. (1972). "Jet Aircraft Engine Noise Reduction". NASA TMX-68131.
823. Yamamoto, K., Brausch, J. F., Balsa, T. F., Janardan, B. A., and Knott, R. (1984). "Experimental Investigation of Shock Cell Noise Reduction for Single Stream Nozzles in Simulated Flight". Final Report NASA-CR-3845. (See also NASA-CR-3846 for Dual Stream Nozzles.)
824. Crouch, R. W., Coughlin, C. L., and Paynter, G. L. (1976). "Nozzle Exit Flow profile Shaping for Jet Noise Reduction". AIAA 76-511.
825. Schwartz, I. R. (1973). "Jet Noise Suppression by Swirling the Jet Flow". AIAA 73-1003.
826. Brooks, J. R., and Woodrow, R. J. (1975). "Silencing an Executive Aircraft". Noise Control Engineering 5:66–74.
827. Grande, E., Brown, D., Sutherland, L., and Tedrick, R. (1973). "Small Turbine Engine Noise Reduction." Vol. II, AFAPL-TR-73-79, Air Force Aero Propulsion Laboratory.
830. Busemann, A. (1955). "The Relation between Minimising Drag and Noise at Supersonic speeds". Proc. Conf. on High-Speed Aeronautics. Polytech. Inst. of Brooklyn.
831. McClean, F. E., and Strout, B. L. (1966). "Design Methods for Minimisation of Sonic Boom Pressure Field Disturbances". J. Ac. Soc. Am. 39:2.
832. Seebass, R., and George, A. R. (1972). "Sonic Boom Minimization". J. Ac. Soc. Am. 51:686–94.
833. Kane, E. J. (1973). "A Study to Determine the Feasibility of a Low-Sonic-Boom Supersonic Aircraft". AIAA 73-1035.
834. Simcox, C. D., Armstrong, R. S., and Atvars, J. (1975). "Recent Advances in Exhaust Systems for Jet Noise Suppression of High Speed Aircraft". AIAA 75-333.
835. Hoch, R. G., and Hawkins, R. (1973). "Recent Studies into Concorde Noise Reduction". AGARD C.P.131, "Noise Mechanisms".
836. Calder, P. H., and Gupta, P. C. (1977). "The Application of New Technology for Performance Improvement and Noise Reduction of Supersonic Transport Aircraft". AIAA & SAW Propulsion Conference, Orlando, Fla.
837. Hiatt, D. L., McKaig, M. B., et al. (1973). "727 Noise Retrofit Feasibility". Three-Part Report No. FAA-RD-72-40.
838. Brooks, J. R., McKinnon, R. A., and Johnson, E. S. (1980). "Results from Flight Noise Tests on a Viper Turbojet Fitted with Ejector-Suppressor Nozzle Systems". AIAA 80-1028.
839. Fitzsimmons, R. D., McKennon, R. A., and Johnson, E. S. (1979). "Flight Test and Tunnel Test Results of the MDC Mechanical Jet Noise Suppressor". NASA CP-2108.
840. Gupta, P. C. (1980). "Advanced Olympus for Next Generation Supersonic Transport". SAE 800732.
841. Anon. (1972). "Concorde – Airport Noise and Silencing Programme". Report to

348   *References*

ICAO by the four Concorde partner companies, issued by British Aircraft
Corporation as report DO/JAH/LG/8904.
845. Kester, J. E. (1974). "Status of the JT8D Refan Noise Reduction Program". Proc.
Inter-noise, 145–50.
846. Wilson, N. J. (1988). "Developing the Rolls-Royce Tay". ASME 88-GT-302.
851. Smith, M. J. T. (1976). "Quietening a Quiet Engine – The RB211 Demonstrator
Programme". SAE 760897.
852. Burdsall, E. A., and Urban, R. H. (1973). "Fan-Compressor Noise: Prediction,
Research and Reduction Studies". FAA-RD-71-73.
853. Rao, G. V. R. (1971). "Study of Non-Radial Stators for Noise Reduction". NASA
CR-1882.
854. Embleton, T. F. W., and Thiessen, G. J. (1970). "Noise Reduction of
Compressors Using Segmented Stator Blades". Canadian Aero. and Space J.
(Nov. 1970):369–73.
855. Lumsdaine, E. (1972). "Development of a Sonic Inlet for Jet Aircraft". InterNoise
72, Proceedings.
856. Klujber, F. (1973). "Development of Sonic Inlets for Turbofan Engines". Journal
of Aircraft, October.
857. Groth, H. W. (1974). "Sonic Inlet Noise Attenuation and Performance with a J-85
Turbojet Engine as a Noise Source". AIAA 74-91.
858. Koch, R. L., Ciskowski, T. M., and Garzon, J. R. (1974). "Turbofan Noise
Reduction Using a Near Sonic Inlet". AIAA Paper 74-1098.
859. Sloan, D., and Farquhar, B. E. (1975). "The Refracting Inlet: A New Concept for
Aircraft Inlet Noise Suppression". ASME 75-GT-71.
860. Sarin, S. L., and Cornelisse, D. A. (1977). "A Novel Concept for Suppression of
Internally Generated Aircraft Engine Noise". AIAA 77-1356.
861. Kazin, S. B., et al. (1973). "Acoustic Testing of 1.5 Pressure Ratio Low Tip Speed
Fan with Casing Tip Bleed (QEP Fan B Scale Model)". NASA CR-120822.
862. Moore, C. J. (1975). "Reduction of Fan Noise by Annulus Boundary layer
Removal". J. S. Vib. 43(4).
863. Karamcheti, K., and Yu, Y. H. (1975). "Aerodynamic Design of a Rotor Blade for
Minimum Noise Radiation". AIAA 74-571.
864. Graham, R. R. (1934). "The Silent Flight of Owls". JRAeS 38:837–43.
865. Hersh, A. S., and Hayden, R. E. (1971). "Aerodynamic Sound Radiation from
Lifting Surfaces with and without Leading Edge Serrations". NASA CR-114370.
866. Kazin, S. B., Paas, J. E., and Minzner, W. R. (1973). "Acoustic Testing of a 1.5
Pressure-Ratio Low-Tipspeed Fan with a Serrated Rotor". NASA CR-12084.
867. Soderman, P. T. (1973). "Leading Edge Serrations Which Reduce the Noise of Low
Speed Rotors". NASA TND-7371.
868. Ko, S. H. (1971). "Sound Attenuation in Lined Rectangular Ducts with Flow and
Its Application to the Reduction of Aircraft Engine Noise". J. Ac. Soc. Am.
50(6):Pt. 1, 1418–32.
869. Thompson, J. R., and Smith, M. J. T. (1970). "Minimum Noise Pod Design".
SAE, 700805.
870. Kazin, S. B., Matta, R. K., Bilwakesh, K. R., et al. (1974). "Core Engine Noise
Control Program: Vol II- Identification of Noise Generation and Suppression
Mechanisms". FAA-RD-74-125.
880. Healy, G. J. (1972). "Investigation of Propeller Vortex Noise including the Effects
of Boundary Layer Control". Lockheed-California Report.
881. Succi, G. P. (1979). "Design of Quiet Efficient Propellers". SAE 790584.
882. Metzger, F. B. (1980). "Progress in Propeller/Prop-Fan Noise Technology". AIAA
80-0856.
883. Bowes, M. A. (1972). "Test and Evaluation of a Quiet Helicopter Configuration
HH-43B". Technical Report 71-31, U.S. Air Mobility Research and Development
Laboratory.

884. Lyon, R. H., Mark, W. D., and Pyle, R. W., Jr. (1973). "Synthesis of Helicopter Rotor Tips for Less Noise". J. Ac. Soc. Am. 53(2):607–18.
885. Magliozzi, B. (1976 and 1979). "V/STOL Rotary Propulsor Noise Prediction and Reduction". FAA-RD-76-49 and FAA-RD-79-107.
890. Kempton, A. J. (1976). "The Ambiguity of Acoustic Sources – A Possibility for Active Control?" J. S. Vib. 48(4).
891. Prydy, R. A., Revel, J. D., Balena, F. J., and Hayward, J. L. (1983). "Evaluation of Interior Noise Control Treatments for High Speed Propfan-Powered Aircraft". AIAA 83-0693.
892. Pope, L. D., Wilby, E. G., Willis, C. M., and Mayes, W. H. (1983). "Aircraft Interior Noise Models: Sidewall Trim, Stiffened Structures and Cabin Acoustics with Floor Partition". J. S. Vib. 89(3).

Noise measurement, analysis and prediction

900. Anon. "Sound System Equipment". IEC 268 and Supplements.
901. Anon. "Sound Level Meters". IEC 651.
902. Anon. "Recommendations for Sound Level Meters". IEC 123.
903. Anon. "Precision Sound Level Meters". IEC 179 and Supplement 179A, "Additional Characteristics for the Measurement of Impulsive Sounds".
904. Anon. "Electro-acoustical Measuring Equipment for Aircraft Noise Certification". IEC 561.
905. Anon. "Description and Measurement of Physical Properties of Sonic Booms". ISO 2249.
906. Anon. "Procedure for Describing Aircraft Noise Heard on the Ground". ISO3891.
907. Anon. "Measurement of Noise inside Aircraft". ISO DIS 5129.
908. Anon. "Type Measurements of Aircraft Interior Noise in Cruise". SAE ARP 1323.
909. Anon. "Measurement of Exterior Noise Produced by Aircraft Auxiliary Power Units (APUs)" SAE ARP 1307.
910. Hassal, J. R., and Zaveri, K. (1971). "Acoustic Noise Measurements". Bruel and Kjaer.
911. Bentley, L. R. (1980). "The Measurement of Aero Gas Turbine Noise". NASA Report PNR-90032.
912. Taniguchi, H. H., and Rasmussen, G. (1979). "Selection and Use of Microphones for Engine and Aircraft Noise Measurements". J. S. Vib. 13:12–20.
913. Anon. "Measurement of Noise from Gas Turbine Engines during Static Operation". SAE AIR 1846.
914. Anon. "Methods of Controlling Distortion of Inlet Airflow during Static Acoustical Tests of Turbofan Engines and Fan Rigs". SAE AIR 1935.
915. Lowrie, B. W., and Newby, D. R. (1977). "The Design and Calibration of a Distortion-Reducing Screen for Fan Noise Testing". AIAA 77-1323.
916. Ginder, R. B., Kenison, R. C., and Smith, A. D. (1979). "Considerations for the Design of Inlet Flow Conditioners for Static Fan Noise Testing". AIAA 79-065.
917. Smith, M. J. T. (1977). "International Aircraft Noise Measurement Procedures: Expensive Acquisition of Poor Quality Data". AIAA 77-1371.
918. Anon. "Practical Methods to Obtain Free-Field Sound Pressure Levels from Acoustical Measurements over Ground Surfaces". SAE AIR 1672.
919. Anon. "Noise Data Sheets; Volume 5, Sound Propagation". ESDU, London.
920. Willshire, W. L., and Nystrom, P. A. (1982). "Investigation of the Effects of Microphone Position and Orientation on Near-Ground Noise Measurements". NASA TP-2004.
921. Smith, M. J. T. (1989). "Bringing Aircraft Noise Testing Down to Earth". European Propulsion Forum, Paper No. 6, RAeS.

922. Payne, R. C., and Miller, G. E. (1984). "A Theoretical Appraisal of the Use of Ground Plane Microphones for Aircraft Noise Measurement". NPL Report Ac. 103.
923. Payne, R. C. (1985). "An Experimental Appraisal of the Use of Ground Plane Microphones for Aircraft Noise Measurement". NPL Report Ac. 104.
924. Pernett, D. F., and Payne, R. C. (1976). "The Effects of Small Variations in the Height of a Microphone above Ground Surface in the Measurement of Aircraft Noise". NPL Ac. 77.
925. Smith, M. J. T. (1986). "Aircraft Noise Measurement – Alternatives to the Standard 1.2m Microphone Height". AIAA 86-1960.
926. Anon. (1988). "Ground-Plane Microphone Configuration for Propeller-Driven Light-Aircraft Noise Measurement". SAE ARP-4055.
927. Brooks, J. R. (1977). "Flight Noise Studies on a Turbojet Using Microphones Mounted on a 450ft Tower". AIAA 77-1325.
928. Anon. "Comparison of Ground Runup and Flyover Noise Levels". SAE AIR 1216.
929. Harper-Bourne, M. (1970). "Optical Measurements of Jet Turbulence". ISVR Memo 398.
930. Fisher, M. J., and Davies, P. O. A. L. (1964). "Correlation Measurements in a Non-Frozen Pattern of Turbulence". J. Fluid Mech. 18:97–116.
931. Kinns, R. (1976). "Binaural Source Location" J. S. Vib. 44.
932. Kinns, R. (1976). "Experiments Using Binaural Source Location". J. S. Vib. 44(2).
933. Flynn, E. O., and Kinns, R. (1976). "Multiplicative Signal Processing for Sound Source Location in Jet Engines". J. S. Vib. 44.
934. Billingsley, J., and Kinns, R. (1976). "The Acoustic Telescope". J. S. Vib. 48(4).
935. Tester, B. J., and Fisher, M. J. (1981). "Engine Noise Source Breakdown: Theory, Simulation and Results". AIAA 81-2040.
936. Fisher, M. J., and Tester, B. J. (1984). "A Review of Source Location/Assessment Techniques". Ten Years of Research; von Karman Institute.
937. Fisher, M. J., Harper-Bourne, M., and Glegg, S. A. L. (1977). "Jet Engine Source Location: The Polar Correlation Technique". J. S. Vib. 51(1).
938. Pack, P. M. W., and Strange, P. J. R. (1987). "Recent Developments in Source Location". AIAA 87-2685.
939. Howell, G., Bradley, A. J., McCormick, M. A., and Brown, J. D. (1984). "De-Dopplerisation and Acoustic Imaging of Aircraft Flyover Measurements". AIAA and NASA Aeroacoustics Conference, Williamsburg, Va.
940. Lowrie, B. W., Morfey, C. L., and Tester, B. J. (1977). "Far Field Methods of Duct Mode Detection for Broadband Noise Source". AIAA 77-1331.
941. Moore, C. J. (1979). "The Measurement of Radial and Circumferential Modes in Annular and Circular Ducts". J. S. Vib. 62(2).
942. Tester, B. J., Cargill, A. M., and Barry, B. (1979). "Fan Duct Mode Detection in the Far Field: Simulation, Measurement and Anaylsis". AIAA 79-0580.
943. Cargill, A. M. (1980). "Fan Noise Source Location from Far Field Measurements". AIAA 80-1054.
944. Holbeche, T. A., and Hazell, A. F. (1981). "A Novel Airborne Technique for Free-Field Measurements of Aircraft Noise above the Flight Path with Application to Noise Shielding Studies". AIAA 81-2028.
945. Anon. "Noise Data Sheets; Volume 1 and 2, General and Noise Estimation". ESDU, London.
946. Anon. "Acoustic Fatigue Data Sheets". ESDU, London.
947. Anon. "Gas Turbine Jet Exhaust Noise Prediction". SAE ARP 876C.
948. Bushell, K. W. (1971). "A Survey of Low Velocity and Coaxial Jet Noise with Application to Prediction" J. S. Vib. 17(2):271–82.
949. Stone, J. R. (1974). "Interim Prediction Method for Jet Noise". NASA TMX-71618.

950. Jaeck, C. L. (1977). "Empirical Jet Noise Predictions for Single and Dual Flow Jets with and without Suppressor Nozzles". Boeing Company Documents D6-42929, 1 and 2.

951. Cocking, B. J. (1977). "A Prediction Method for the Effects of Flight on Subsonic Jet Noise". J. S. Vib. 53(3):435–53.

952. Stone, J. R. (1979). "An Improved Method for Predicting the Effects of Flight on Jet Mixing Noise". NASA TMX 79-155.

953. Bushell, K. W. (1975). "Measurement and Prediction of Jet Noise in Flight". AIAA 75-461.

954. Bryce, W. D. (1984). "The Prediction of Static-to-Flight Changes in Jet Noise". AIAA 84-2358.

955. Lu, H. Y. (1986). "An Empirical Model for Prediction of Coaxial Jet Noise in Ambient Flow". AIAA 86-1912.

956. Anon. "Gas Turbine Co-axial Exhaust Flow Noise Prediction". SAE AIR 1905.

957. Pao, S. P. (1979). "A Correlation of Mixing Noise from Coannular Jets with Inverted Velocity Profiles". NASA TP-1301.

958. Russell, J. W. (1979). "A Method for Predicting the Noise Levels of Coannular Jets with Inverted Velocity Profiles". NASA CR-3176.

959. Stone, J. R. (1977). "An Empirical Model for Inverted-Velocity-Profile Jet Noise Prediction". NASA TMX-73552.

960. Green, A. R. "Development of a Buzz-Saw Noise Prediction Method". ESDU Data Sheet, London.

961. Benzakein, M. J., and Morgan, W. R. (1969). "Analytical Prediction of Fan/ Compressor Noise". ASME 69-WA/GT-10.

962. Heidman, M. F. (1976 and 1979 Update). "Interim Prediction Procedure for Fan and Compressor Source Noise". NASA TMX-71763.

963. Tyler, J. M., and Sofrin, T. B. (1961). "Axial Flow Compressor Noise Studies". Trans. SAE, pp. 309–32.

964. Smith, M. J. T., and House, M. E. (1967). "Internally Generated Noise from Gas Turbine Engines – Measurement and Prediction." ASME J. Eng. and Power 89.

965. Philpot, M. G. (1975). "The Role of Rotor Blockage in the Propagation of Fan Noise Interaction Tones". AIAA 75-447.

966. Kershaw, R. J., and House, M. E. (1982). "Sound-Absorbent Duct Design". Chapter 21 of Ref. 210.

967. Mathews, C. C., Nagel, R. T., and Kester, J. D. (1975). "Review of Theory and Methods for Turbine Noise Prediction". AIAA 75-540.

968. Krejsa, E. A., and Valerino, M. F. (1977). "Interim Prediction Method for Turbine Noise". NASA TMX-73566.

969. Matta, R. K., Sandusky, G. T., and Doyle, V. L. (1977). "GE Core Engine Noise Investigation – Low Emission Engines". FAA RD-77-4.

970. Anon. "Prediction Procedure for Near-field and Far-field Propeller Noise". SAE AIR 1407.

971. Farassat, F. (1983). "Linear Acoustic Formulas for Calculation of Rotating Blade Noise". AIAA J. 19(9):1122–30.

972. Farassat, F., and Succi, G. P. (1983). "The Prediction of Helicopter Rotor Discrete Frequency Noise". Vertica 309–20.

973. Pegg, R. J. (1979). "A Summary and Evaluation of Semi-empirical Methods for the Prediction of Helicopter Rotor Noise". NASA TM-80200.

974. Farassat, F. (1975). "Theory of Noise Generation from Moving Bodies with Application to Helicopter Rotors". NASA TR-451.

975. Ho, P. Y., and Tedrick, R. N. (1972). "Combustion Noise Prediction Techniques for Small Gas Turbine Engines" Proc. InterNoise.

976. Emmerling, J. J., Kazin, S. B., Matta, R. K., et al. (1974 and 1976). "Core Engine Noise Control Program: Vol. III – Prediction Methods". FAA-RD-74-125 and Supplement.
977. Mathews, D. C., and Rekos, N. F., Jr. (1976). "Direct Combustion Generated Noise in Turbopropulsion Systems – Prediction and Measurement". AIAA 76-579.
978. Ho, P. Y., and Doyle, V. L. (1979). "Combustion Noise Prediction Update". AIAA 79-0588.
979. Huff, R. G., Clark, B. J., and Dorsch, R. G. (1974). "Interim Prediction Method for Low Frequency Core Engine Noise". NASA TMX-71627.
980. Fink, M. R. (1979). "Noise Component Method for Airframe Noise". Journal of Aircraft" 16:659–65.
981. Hardin, J. C., Fratello, D. J., Hayden, R. E., Kadman, Y., and Africk, S. (1975). "Prediction of Airframe Noise". NASA TND-7821.
982. Bliss, D. B., and Hayden, R. E. (1976). "Landing-Gear and Cavity Noise Prediction". NASA CR-2714.
983. Dorsch, R. G., Clark, B. J., and Reseshotko, M. (1975). "Interim Prediction Method for Externally Blown Flap Noise". NASA TM X-71768.
985. Raney, J. P., Padula, S. L., and Zorumski, W. E. (1981). "NASA Progress in Aircraft Noise Prediction". NASA TM-81915.
986. Zorumski, W. E. (1982 and 1986). "Aircraft Noise Prediction Program. Theoretical Manual". NASA TM-83199, (3 Pts.).
987. Pao, S. P., Wenzel, A. R., and Oncley, P. B. (1978). "Prediction of Ground Effects on Aircraft Noise". NASA TP-1104.
988. Dunn, D. G., and Peart, N. A. (1973). "Aircraft Noise Source and Contour Estimation". NASA CR-114649.
989. Galloway, W. J. (1980). "Airport Noise Contour Predictions – Improving Their Accuracy". Proc. Inter Noise.
990. Anon. (1988). "Recommended Method for Computing Noise Contours around Airports". ICAO Circular, 205-AN/1/25.
991. Anon. "Procedure for the Calculation of Airplane Noise in the Vicinity of Airports". SAE AIR 1845.
992. Anon. (1986). "Standard Method of Computing Noise Contours around Civil Airports". European Civil Aviation Conference, Paris, Doc. No. 29.
993. Anon. (1982). "The Consequences of Aircraft Noise Abatement Regulations (CANAR-Airport Noise Contour and Single Event Footprint Calculation)", and Information Bulletin No. 1, Commission of the European Communities.
994. Anon. (1982). "Federal Aviation Administration Integrated Noise Model-Version 3 Users Guide". U.S. Dept. of Transportation Report FAA-EE-81-17.
995. Anon. "Prediction Method for Lateral Attenuation of Airplane Noise during Take-off and Landing". SAE AIR 1751.
996. Anon. "Noise Data Sheets Volume 4, Sound Propagation 2". ESDU, London.
997. Speakman, J. D. (1980). "NOISEMAP – The USAF's Computer Program for Predicting Noise Exposure around an Airport". Proc. Inter Noise.
998. Davies, L. I. C. (1982). "Ground-Roll Noise Modelling in the DORA Noise and Number Index Computer Model and Measurements at Heathrow Airport", Civil Aviation Authority, Report DORA 8204.
999. Svane, C. (1987). "Flight Effect on Aircraft Noise during the Ground Roll". Danish Acoustical Institute Report 138.

Index